Integer Programming

Facets, Subadditivity,
and Duality
for Group and Semi-group
Problems

ELLIS L. JOHNSON
IBM Thomas J. Watson Research Center

SOCIETY for INDUSTRIAL and
APPLIED MATHEMATICS • 1980

PHILADELPHIA, PENNSYLVANIA 19103

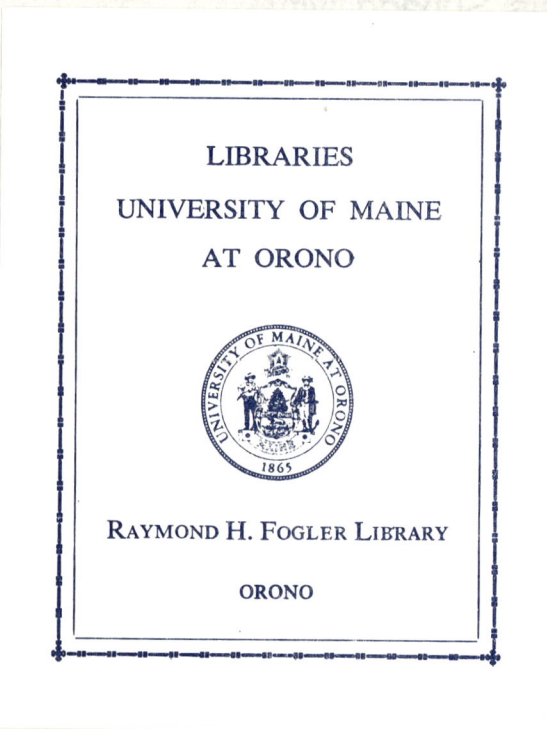

Copyright © 1980 by Society for Industrial and Applied Mathematics.

Library of Congress Catalog Card Number: 79-93152.

ISBN: 0-89871-162-2.

Printed in England for the Society for Industrial and Applied Mathematics by J. W. Arrowsmith Ltd., Winterstoke Road, Bristol BS3 2NT, England.

Contents

Foreword . . . . *v*
Preface . . . . *vii*

Chapter I
INTEGER PROGRAMMING
    1. The problem 1
    2. The importance of integer programming 1
    3. Early developments 1
    4. Strong linear programming relaxations 3
    5. Combinatorial polyhedra 4
    6. Blocking pairs of polyhedra 5
    7. Group and semigroup problems 7

Chapter II
CUTS, KNAPSACKS, AND A CYCLIC GROUP PROBLEM   11
    1. Introduction 11
    2. Gomory's fractional cut 11
    3. The cutting stock problem 13
    4. The knapsack problem 14

Chapter III
FINITE ABELIAN GROUPS   17
    1. Cyclic groups 17
    2. The cyclic group problem 18
    3. Finite Abelian groups 19

Chapter IV
GOMORY'S CORNER POLYHEDRA   23
    1. The group problem 23
    2. The asymptotic theorem 25
    3. Corner polyhedra 26

Chapter V
BLOCKING POLYHEDRA AND MASTER GROUP PROBLEMS   29
    1. Blocking pairs of polyhedra 29
    2. Master group problems 30
    3. Facets of master group problems 31
    4. Automorphisms and homomorphisms 32

Chapter VI
ARAOZ'S SEMIGROUP PROBLEM   35
1. Semigroups 35
2. The semigroup problem 36
3. Master semigroup problems 37

Chapter VII
BLOCKERS AND POLARS FOR MASTER SEMIGROUP PROBLEMS   41
1. The blocker of master semigroup problems 41
2. The polar of master semigroup problems 43

Chapter VIII
SUBADDITIVE AND MINIMAL VALID INEQUALITIES   48
1. The subadditive cone 48
2. Subadditive valid inequalities 48
3. Minimal valid inequalities 50
4. Valid equations 53
5. Homogeneous valid inequalities 55

Chapter IX
SUBADDITIVE CHARACTERIZATIONS   57
1. The blockers $\mathcal{B}(\mathcal{G}, b)$ 57
2. Valid inequalities 59

Chapter X
DUALITY   63
1. A dual problem for master problems 63
2. Semigroup problems for general $\mathcal{N}$ 64
3. Lifting procedures 65

Foreword

The material contained in this monograph formed the basis for a one week Regional Research Conference on Integer Programming. The conference, sponsored by the National Science Foundation, was held at the State University of New York at Buffalo, June 5–9, 1978. Professor Subhash C. Narula (then at SUNY/Buffalo and now at Rensselaer Polytechnic Institute) and I conceived the conference in 1975. We asked Dr. Ellis Johnson if he would like to present lectures that would form a basis for a conference. He responded enthusiastically, and with the help of the National Science Foundation we were able to schedule the conference for June, 1978.

The conference was most successful. We had an excellent group of participants, an outstanding lecturer, and interesting material. Such a pleasant outcome is due in large measure to the principal speaker and author of the monograph: Dr. Ellis Johnson; the sponsor: The National Science Foundation and its representative, Dr. William H. Pell; and the coorganizers: Professor Narula and myself.

I also wish to thank the behind the scene participants at the State University of New York at Buffalo for their support: Dr. Joseph A. Alutto, Dean of the School of Management; Mr. Nelson K. Upton, Administrator, The Center for Management Development in the School of Management; the Food Service and Housing Office; and my secretary, Mrs. Marilyn Viau, for her extra efforts in helping to overcome all the problems arising in organizing and running the conference.

It is, therefore, with great pride that I write this Foreword in the hope that this volume will become an important part of the integer programming literature.

STANLEY ZIONTS
State University of New York at Buffalo

Preface

These lecture notes draw upon the work of several important contributors; in particular I would like to mention Ralph Gomory, Ray Fulkerson, and Julian Araoz. Other people who have had a strong influence on my approach and view of integer programming are: Jack Edmonds, who among other things first suggested that the subadditive approach was directly applicable to integer programs; Alan Hoffman, who has patiently listened to and suggested improvements to this entire development over a period of ten years; and Claude Burdet, who first suggested that I "lift" my subadditive functions rather than using tables of facets.

The conference itself was very pleasant, and I wish especially to thank Stan Zionts for thinking of holding it in the first place and then doing such an excellent job, along with Subhash Narula, in carrying it out so that everyone had such a worthwhile week. They were also active participants in the technical discussions throughout the week. The facilities of SUNY/Buffalo were excellent, the social side of the week was well conceived and pleasant, and everyone seemed to enjoy the conference.

The participants were a lively group, making lecturing a challenge and a pleasure. We enjoyed the lecture by Tom Baker on the integer programming work at Exxon and the lecture by Bob Bixby on his work with Bill Cunningham in converting linear programs to network flow problems. Maurice Queyranne converted a conjecture in the original lecture notes into a theorem (Theorem 2 of Chapter VI). Several participants contributed excellent ideas during problem sessions and discussions. Bob Jeroslow was an ideal choice by NSF as observer and contributed his own insights. In general, the participants were lively, interested, and a pleasure to be with.

Finally, let me thank the National Science Foundation and the Conference Board of the Mathematical Sciences for setting up these regional conferences and for supporting this one in particular. From my point of view as a lecturer, the conference was a most valuable stimulus, which happened to come at a time when I needed the discipline imposed by the necessity of organizing these lectures. It was a pleasure to immerse myself in the brilliant work of Gomory, Fulkerson, and Araoz and to tackle the task of pulling together this material.

During the past five years, because of a combination of circumstances, I had not continued developing this approach at the same pace as earlier. However, the conference and my preparation of these notes has revived my enthusiasm, and I have derived several important, new results which I am still in the process of working out. I will certainly remain indebted to this conference for giving me a strong push. Thus, it is with sincere gratitude that I thank Stan, the NSF, the CBMS, and all of those who came to Buffalo.

ELLIS L. JOHNSON
Yorktown Heights, March 1979

CHAPTER I

Integer Programming

1. The problem. The general mixed *integer programming problem* is to minimize $z = cx$ subject to

(1) $\qquad x_j \text{ integer } j \in J \subseteq \{1, \cdots, n\}, \qquad J \neq \emptyset,$

(2) $\qquad x_j \geq 0, \quad j = 1, \cdots, n,$

(3) $\qquad Ax = b.$

An integer programming problem is a *pure integer program* if $J = \{1, \cdots, n\}$. The *linear programming relaxation* of the integer programming problem is the corresponding linear program with constraints (2) and (3) imposed, but not (1).

2. The importance of integer programming. Shortly after the development of his simplex algorithm for linear programs, Dantzig [6] pointed out the significance of solving integer programming problems. Many problems involving nonconvex regions or functions can be converted into integer programs, e.g. linear programs with separable, piecewise linear, but nonconvex objective functions. The challenge raised by Dantzig was to develop effective procedures, such as his simplex method was proving to be for linear programs, to handle these more difficult integer programs.

Today, there are codes, including commercial codes, which effectively solve a small, but important, class of integer programs. This class is not well defined but can be said to include integer programs with linear programming relaxations which may be large, i.e. up to several thousand rows and columns, but with a small number of integer variables $x_j, j \in J$, or a strong linear programming relaxation. In § 4, the meaning of "strong" in this context is discussed.

An example of use of integer programming codes is a production distribution model for deciding which car lines to make at which plants. Estimates of two to three dollars in savings for each car produced have been given [23]. With yearly production in the millions, one quickly grasps the importance of integer programming. It is fair to say that using large computers and large linear programs of a global or strategic nature, integer programming codes are able to consider an enormous number of possibilities in order to improve upon manual planning.

3. Early developments. An important and interesting class of linear programs was seen by Dantzig [5] to always have integer answers. This class was network flow problems [13]. Any combinatorial optimization problem which can be formulated as a network flow problem can be solved by solving a linear program; not only a linear program but one with special structure and faster methods. The work of Dantzig and Hoffman [8], Fulkerson [14], [15], Hoffman [25], and Hoffman and Kuhn [26] led to classifying several problems (e.g. assignment, incomparable nodes in a partial order) and associated duality theorems (Konig–Etchevary and Dilworth's theorems) as instances of network flow problems for which linear programming duality gives interesting, frequently well known, theorems.

The work of Dantzig, Fulkerson, and Johnson [7], on the traveling salesman problem, used linear programming to solve problems which were integer programs not of the network flow type. This ground breaking work took a difficult problem, formulated a huge linear programming relaxation, and used both cutting plane and branch and bound methods to try to find optimum integer answers. The problem is, given a graph with edges (undirected) $[i, j] \in E$, find an incidence vector $(x_{ij}, [i, j] \in E)$ so as to minimize $\sum c_{ij} x_{ij}$. The linear programming relaxation is

$$0 \leq x_{ij} \leq 1, \quad i = 1, \cdots, n, \quad j = 1, \cdots, n, \quad i \neq j;$$

$$\sum_i x_{ij} = 2, \quad \text{all } j = 1, \cdots, n;$$

$$\sum_{i \in S, j \notin S} x_{ij} \geq 2, \quad \text{all proper subsets } S \subseteq \{1, \cdots, n\}.$$

There are a huge number of these latter constraints, so they are adjoined only selectively when violated. Despite the fact that there are smaller linear programming relaxations, this one continues to be used and is the basis for the work of Held and Karp [24], Miliotis [33], and Hong and Padberg [27]. The reason it is used is that it is a "strong" relaxation, in practice, as will be discussed in the next section.

In the late 1950's, the cutting plane work of Gomory [19] was being developed. We discuss cutting planes in the next section. At the same time, Land and Doig [31] and Beale [3] were developing a simpler, but useful, method: branch and bound. This method solves the linear programming relaxation, and then creates two or more problems by branching. For example, if x_j is a variable which should be zero or one and takes on a fractional value in the linear programming optimum, two problems are created with $x_j = 0$ in one and $x_j = 1$ in the other. The bounds come from the objective function values of the linear programming relaxations.

The main reason that branch and bound is used in commercial codes is that it is ideal for mixed integer problems. Success depends upon factors such as: number of integer variables, strength of the linear programming relaxation, and whether or not there are many integer answers.

For pure integer problems, cutting plane methods sometime work well and sometime not at all. Branch and bound can spend too much time solving linear programs. Enumeration [2] was a name originally used for methods which solved no linear programs and kept integer variables at some trial values. With use of linear programs in enumeration, the boundaries become unclear. See Garfinkel and Nemhauser [17] and Geoffrion and Marsten [18] for general discussions. Enumeration methods can still be said to rely less heavily on linear programming and to employ feasibility and logical tests [22], [34] and efficient data structures outside of the linear programming tableau in order to quickly search large numbers of possibilities.

4. Strong linear programming relaxations. The considerations given here turn out to be very important practically. Any method that solves linear programs as part of a method to try to solve integer programs will profit from a better linear programming formulation.

In § 1, we defined the integer programming problem and its linear programming relaxation. Correspondingly, define the *convex hull of integer solutions* to be

$$\mathscr{P}_I = \text{conv } \{x | x_j \geq 0, j = 1, \cdots, n; x_j \text{ integer } j \in J; \text{ and } Ax = b\}$$

and the linear programming polyhedron to be

$$\mathscr{P}_L = \{x | x_j \geq 0, j = 1, \cdots, n; \text{ and } Ax = b\}.$$

Clearly, $\mathscr{P}_I \subseteq \mathscr{P}_L$.

There are many linear programs giving the same $\mathscr{P}_I$ but different $\mathscr{P}_L$'s. We illustrate with an example.

Example 1. Consider the constraints:

$$x_1 + x_2 \leq 2y, \quad 0 \leq x_j \leq 1, \quad j = 1, 2, \text{ and } y = 0 \text{ or } 1.$$

The two polyhedra $\mathscr{P}_I$ and $\mathscr{P}_L$ are shown in Fig. 1. Here,

$$\mathscr{P}_I = \{(x_1, x_2, y) | 0 \leq x_1 \leq y, 0 \leq x_2 \leq y, 0 \leq y \leq 1\}.$$

If we had originally stated the problem as having constraints

$$0 \leq x_1 \leq y, \quad 0 \leq x_2 \leq y, \quad y = 0 \text{ or } 1,$$

then the linear programming relaxation would have had $\mathscr{P}_L = \mathscr{P}_I$.

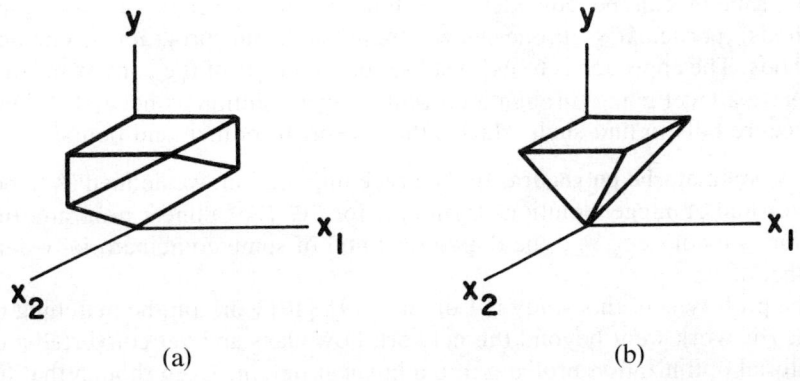

Fig. 1

The strongest linear programming relaxation is one that gives $\mathscr{P}_L = \mathscr{P}_1$. This linear program exists, in fact, provided A is a rational matrix. In practice it is usually not known and even if it were it would have many constraints. In any case, the inclusion ordering on $\mathscr{P}_L$ gives a partial ordering on all possible linear programming relaxations. In Example 1, an even weaker linear programming relaxation than the one given is

$$0 \leq x_1 \leq 1, \quad 0 \leq x_2 \leq 1, \quad x_1 + x_2 \leq 3y, \quad 0 \leq y \leq 1.$$

In that example, any linear program which imposes $0 \leq y \leq 1$ and whose feasible region for $y = 0$ is $x_1 = x_2 = 0$ and for $y = 1$ is $0 \leq x_1 \leq 1, 0 \leq x_2 \leq 1$, will give a legitimate linear programming relaxation.

When we speak of a strong linear programming relaxation we mean, loosely speaking, one whose objective function value is close to the integer optimum objective value, i.e. one so that $\mathscr{P}_L$ is not much larger than $\mathscr{P}_1$ in any direction.

Cutting plane methods try to strengthen the linear programming relaxation so that the optimum linear programming solution is integer; that is, they try to add inequalities to $\mathscr{P}_L$ so that its optimum belongs to $\mathscr{P}_1$.

Cutting plane methods have been mainly developed for pure integer problems. This fact alone would account for commercial codes using branch and bound, rather than cutting plane methods. The main reported large scale use of cutting planes is that of Glenn Martin [32] on airline crew scheduling problems. His problems are particular pure integer problems with a large part being set covering: $Ax \geq b, x = 0$ or 1, where A is a matrix of 0's and 1's and b is a vector of 1's.

Among pure integer problems, there are some for which a particular cutting plane method will break down completely. It is not understood which class of problems are amenable to solution with a given cutting plane method. Nevertheless, it is a very good idea to strengthen the linear program before using other techniques, such as branch and bound or enumeration. Thus, one cannot rule out cutting plane methods, especially when judiciously combined with other methods [30].

The computational work of Crowder and Padberg [4] on the traveling salesman problem can be considered within the broad scope of cutting plane methods; particularly if one views them as linear program strengthening methods. The approach is to use a subset of the facets of the convex hull and to generate a facet going through a current integer solution as needed. When the procedure fails to find such a facet, they resort to branch and bound.

5. Combinatorial polyhedra. In the preceding section, we defined $\mathscr{P}_1$ to be the convex hull of integer solutions (x_j integer for $j \in J$) of a linear program. In this section, we consider $\mathscr{P}_1$ to be the convex hull of some combinatorially defined polyhedra.

The prototype of this study is Edmonds [9], [10] work on the matching polytope. His work went beyond the network flow class and yet converted a combinatorial optimization problem into a linear program. Even though that linear program has an enormous number of inequalities, he described a good (polynomially bounded) algorithm for solving the matching problem.

Given an undirected graph with vertices V and edges E, the matching problem is to find subset M of edges so that no two edges of M meet the same vertex and maximize $\sum_{e \in M} c(e)$ over all such M, where $(c(e), e \in E)$ is given as the objective function.

The matching problem is the integer program

$$\text{maximize} \sum_{e \in E} c(e)x(e) \text{ subject to}$$

$$x(e) \geq 0 \text{ and integer}, e \in E, \text{ and}$$

$$\sum_{e \text{ meets } v} x(e) \leq 1.$$

However, the *matching polytope* $\mathcal{P}_1$ is

$\mathcal{P}_1 = \text{conv } \{(x(e), e \in E) | x(e) = 0 \text{ or } 1 \text{ and } M = \{e | x(e) = 1\} \text{ is a matching}\}.$

There are various linear programming relaxations of the matching problem, but the focus here is to begin with $\mathcal{P}_1$ and try to learn as much as possible about it. In particular, we would like to characterize all of its facets. This characterization was given by Edmonds as

$$\sum_{e \in D(S)} x(e) \leq \frac{|S| - 1}{2},$$

where S is a subset of nodes of odd cardinality and $D(S)$ is the subset of edges $e \in E$ with both ends of e meeting nodes of S. That is, $\mathcal{P}_1$ is equal to the solution set

$$\left\{(x(e), e \in E) | x(e) \geq 0, \sum_{e \text{ meets } v} x(e) \leq 1, \text{ and } \sum_{e \in D(S)} x(e) \leq \frac{|S| - 1}{2}\right\}.$$

When we refer to the matching polytope, we mean the convex hull of incidence vectors of sets M which are matchings. When we say the linear characterization of the matching polytope, we mean the description of all additional inequalities to convert the problem to a linear program as Edmonds did for the matching problem. It is more difficult to answer which are actually facets (see Edmonds and Pulleyblank [12] for matchings). However, Edmonds' algorithm for finding optimum matchings did not depend on knowing which were actually facets. In any case, the facets were among the set given.

Other polytopes whose linear characterizations have been found are: the polytope of subsets M of E which are independent in two matroids M_1 and M_2 over E [11], and the polytope of collections of edges which are of even degree at certain nodes and of odd degree at the other nodes [17].

6. Blocking pairs of polyhedra. The results in this section are due to Fulkerson [16]. We return to this subject in Chapter V and give related results. Here, we stay with Fulkerson's original development.

Let A be an $m \times n$ nonnegative matrix and consider the polyhedron

$$\mathcal{P} = \{x \in \mathcal{R}^n | Ax \geq 1, x \geq 0\}.$$

Then, $\mathcal{P}$ is the vector sum of the convex hull of its extreme points and the nonnegative orthant $\mathcal{R}^n_+$:

$$\mathcal{P} = \text{conv}(V) + \mathcal{R}^n_+.$$

Define the *blocking polyhedron* $\mathcal{B}(\mathcal{P})$ of $\mathcal{P}$ to be

$$\mathcal{B}(\mathcal{P}) = \{x^* \geq 0 | x^* \cdot x \geq 1 \text{ for all } x \in \mathcal{P}\}.$$

The blocking polyhedron has a similar representation

$$\mathcal{B}(\mathcal{P}) = \text{conv}(V^*) + \mathcal{R}^n_+,$$

where V^* is the set of vertices of $\mathcal{B}(\mathcal{P})$.

There is a great symmetry here (see [16, Thm. 2.1]). The vertices $v^* \in V^*$ of $\mathcal{B}(\mathcal{P})$ are among the rows of A and are those rows such that $A_i \cdot x \geq 1$ is needed to define $\mathcal{P}$; i.e., the facets of $\mathcal{P}$ are the inequalities which cannot be deduced from a nonnegative combination of the other inequalities $A_k \cdot x \geq 1$, $k \neq i$. We could, in fact, start with such a matrix A and then

$$\mathcal{P} = \{x \in \mathcal{R}^n | Ax \geq 1, x \geq 0\}$$
$$= \text{conv}(A^*) + \mathcal{R}^n_+ \quad \text{and}$$
$$\mathcal{B}(\mathcal{P}) = \{x^* \in \mathcal{R}^n | A^* x^* \geq 1, x^* \geq 0\}$$
$$= \text{conv}(A) + \mathcal{R}^n_+,$$

where A^* is the matrix whose rows are vertices of $\mathcal{P}$ and where $\text{conv}(A^*)$ denotes the convex hull of the rows of A^*. Thus, the facets of $\mathcal{P}$, other than $x \geq 0$, are the vertices of $\mathcal{B}(\mathcal{P})$; and the facets of $\mathcal{B}(\mathcal{P})$, other than $x^* \geq 0$, are the vertices of $\mathcal{P}$. Furthermore,

$$\mathcal{B}(\mathcal{B}(\mathcal{P})) = \mathcal{P}.$$

The matrices A and A^* are called a *blocking pair of matrices*.

Fulkerson showed other results (see [16, Thm. 3.1]) which we do not directly develop but leave as part of some exercises.

A particularly appealing special case of blocking pairs is when A and A^* are both 0 1 matrices. An example is when A is the set of source to sink chains (simple, directed paths) in a directed network and A^* is the set of directed cuts separating source and sink. In this case, the rows of A are incidence vectors of chains, and the chains form a clutter (no chain is a proper subset of any other chain). The rows of A and A^* correspond to *blocking clutters*, i.e., every chain crosses at least one cut and every cut is met by at least one chain. However, for A and A^* to be a blocking pair in the sense described earlier, it is necessary and sufficient that either $Ax \geq 1$, $x \geq 0$ has only 0 1 vertices or $A^* x^* \geq 1$, $x^* \geq 0$ has only 0 1 vertices, and if either holds, then so does the other.

7. Group and semigroup problems. The development of Gomory's group approach is traced in Chapter II. The main results are in Chapter IV. Let us address, here, the general question: What has this approach done for us, and where, if anywhere, is it going?

The first thing one can point to from the group problem is a descriptive theorem about what must be done to the solution of the linear programming relaxation in order to get an optimum integer answer in certain cases (see Chapter IV, §2). It is a result which should be appreciated for what it says, namely how to change the linear programming solution to an integer answer, and not so much deprecated for what it does not say, namely what happens in case the conditions of the theorem do not hold. Thus, the theorem in Chapter IV, §2 can be viewed as a nicely descriptive theorem without getting tied up with how to use it to solve integer programming problems.

The second result of Gomory's group work [21] which has proven to be useful is his subadditive characterization of facets (see Chapter IV, §3). This result has theoretical content [21] and has duality (see Chapter X, §1) and algorithmic (see Chapter X, §3) implications. It has been generalized to semigroup problems by Araoz [1] (see Chapter VI), to the mixed integer problem [28], and to general additive systems [29]. In fact, this result can well be said to be the central, underlying result of this volume, and this volume represents only one part of the development, i.e. we do not even mention the mixed integer problem, and the algorithmic possibilities are only touched on.

What we do attempt to accomplish is to expose the initial developments, relate them to Fulkerson's blocking theory of polyhedra, and show how Araoz's extensions lead to being able to apply the group results directly to certain pure integer programming problems.

REFERENCES

[1] J. ARAOZ, *Polyhedral neopolarities,* Ph.D. thesis, Dept. of Computer Sciences and Applied Analysis, Univ. of Waterloo, Waterloo, Ontario, December 1973.

[2] E. BALAS, *An additive algorithm for solving linear programs with zero one variables,* Operations Res., 13 (1965), pp. 517–546.

[3] E. M. L. BEALE, *A method of solving linear programming problems when some but not all of the variables must take integral values,* Statistical Techniques Research Group, Princeton Univ. Princeton, NJ, 1958.

[4] H. P. CROWDER AND M. W. PADBERG, *Solving large-scale symmetric travelling salesman problems to optimality,* IBM Thomas J. Watson Research Center, Yorktown Heights, NY, 1979.

[5] G. B. DANTZIG, *Application of the simplex method to a transportation problem,* Activity Analysis of Production and Allocation, T. C. Koopmans, ed., John Wiley, New York, 1951, pp. 339–347.

[6] ———, *On the significance of solving linear programming problems with some integer variables,* Econometrica, 28 (1960), pp. 30–44.

[7] G. B. DANTZIG, D. R. FULKERSON AND S. M. JOHNSON, *Solution of a large-scale travelling-salesman problem,* Operations Research, 2 (1954), pp. 393–410.

[8] G. B. DANTZIG AND A. J. HOFFMAN, *Dilworth's theorem on partially ordered sets,* Linear

Inequalities and Related Systems, Annals of Mathematics Study 38, Princeton University Press, Princeton, NJ, 1956, pp. 207–214.

[9] J. EDMONDS, *Paths, trees, and flowers*, Canad. J. Math., 17 (1965), pp. 449–467.

[10] ———, *Maximum matching and a polyhedron with 0, 1 vertices*, J. Res. Nat. Bur. Standards Sect. B, 69 (1965), pp. 125–130.

[11] ———, *Submodular functions, matroids, and certain polyhedra*, Combinatorial Structures and Their Applications, R. Guy et al., eds., Gordon and Breach, New York, 1970, pp. 69–88.

[12] J. EDMONDS AND W. R. PULLEYBLANK, *Facets of 1-matching polyhedra*, Rep. CORR 73-3 (1973), Dept. of Combinatorics and Optimization, Univ. of Waterloo, Waterloo, Ontario.

[13] L. R. FORD, JR. AND D. R. FULKERSON, *Flows in Networks*, Princeton University Press, Princeton, NJ, 1962.

[14] D. R. FULKERSON, *Note on Dilworth's decomposition theorem for partially ordered sets*, Proc. Amer. Math. Soc., 7 (1956), pp. 701–702.

[15] ———, *A network flow feasibility theorem and combinatorial applications*, Canad. J. Math., 11 (1959), pp. 440–451.

[16] ———, *Blocking polyhedra*, Graph Theory and Its Applications, B. Harris, ed., Academic Press, New York, 1970, pp. 93–112.

[17] R. S. GARFINKEL AND G. L. NEMHAUSER, *Integer Programming*, John Wiley, New York, 1972.

[18] A. M. GEOFFRION AND R. E. MARSTEN, *Integer programming: A framework and state-of-the-art survey*, Management Sci., 18 (1972), pp. 465–491.

[19] R. E. GOMORY, *An algorithm for integer solutions to linear programs*, Recent Advances in Mathematical Programming, R. L. Graves and P. Wolfe, eds., McGraw-Hill, New York, 1963, pp. 269–302.

[20] ———, *Some polyhedra related to combinatorial problems*, Linear Algebra and Appl., 2 (1969), pp. 451–558.

[21] R. E. GOMORY AND E. L. JOHNSON, *Some continuous functions related to corner polyhedra*, Math. Programming, 3 (1972), pp. 23–85 and pp. 359–389.

[22] P. L. HAMMER, *Boolean procedures for bivalent programming*, Mathematical Programming in Theory and Practice, P. L. Hammer and G. Zowtendijk, eds., North-Holland, Amsterdam, 1974, pp. 311–363.

[23] P. HAMMER, E. L. JOHNSON AND B. KORTE, eds., Ann. Discrete Math., 5 (1979), Conclusive remarks, § 4.1.

[24] M. HELD AND R. M. KARP, *The traveling-salesman problem and minimum spanning trees: Part II*, Math. Programming, 1 (1971), pp. 6–25.

[25] A. J. HOFFMAN, *Some recent applications of the theory of linear inequalities to extremal combinatorial analysis*, Proc. Symposia on Applied Math, 10 (1960).

[26] A. J. HOFFMAN AND H. KUHN, *On systems of distinct representatives*, Linear Inequalities and Related Systems, Annals of Mathematics Study 38, Princeton University Press, Princeton, NJ, 1956, pp. 199–206.

[27] S. HONG AND M. W. PADBERG, *On the symmetric travelling salesman problem: A computational study*, RC 6720, IBM Thomas J. Watson Research Center, Yorktown Heights, NY, 1977.

[28] E. L. JOHNSON, *On the group problem for mixed integer programming*, Math. Programming Stud., 2 (1974), pp. 137–179.

[29] ———, *On the generality of the subadditive characterization of facets*, RC 7521, IBM Thomas J. Watson Research Center, Yorktown Heights, NY, 1979; to appear in Journal of Algorithms.

[30] E. L. JOHNSON AND U. SUHL, *Experiments in integer programming*, RC 7831 IBM Thomas J. Watson Research Center, Yorktown Heights, NY, 1979; to appear in Discrete Applied Mathematics.

[31] A. H. LAND AND A. G. DOIG, *An automatic method of solving discrete programming problems*, Econometrica, 28 (1960), pp. 497–520. (First appeared as part of a report of the London School of Economics, 1957.)

[32] G. T. MARTIN, *An accelerated Euclidean algorithm for integer linear programming*, Recent Advances in Mathematical Programming, R. L. Graves and P. Wolfe, eds., McGraw-Hill, New York, 1963, pp. 311–318.
[33] P. MILIOTIS, *Using cutting planes to solve the symmetric traveling salesman problem*, Math. Programming, 15 (1978), pp. 177–188.
[34] K. SPIELBERG, *Minimal preferred variable reduction for zero one programming*, Rep. 320-3024, IBM Phila. Science Center, Philadelphia, PA, 1973.

CHAPTER II

Cuts, Knapsacks, and a Cyclic Group Problem

1. Introduction. Two lines of development led to focusing on a group relaxation of integer programming and on a characterization of facets of a corresponding polyhedron. The first of these developments was Gomory's fractional cutting plane method [3]. He exhibits a family of cuts with a striking "cyclic group" character. However, the group ideas were not explicitly developed at that stage.

The second development grew out of work by Gilmore and Gomory on the cutting stock problem [1]. Their column generation scheme required solving a knapsack problem [2] for each column generated. A certain periodicity there led Gomory to his earlier cutting plane work and led to his asymptotic theorem [4] (see Chapter IV, § 2). A corresponding polytope is the subject of the paper [5], but the subadditive characterization of facets only appears in [6]. The two papers [5] and [6] are for more general groups than cyclic groups, but in [7] the development returns to cyclic groups. The proof to be given here is closest to that in [7]. For this chapter, we restrict ourselves to finite cyclic groups.

2. Gomory's fractional cut. Begin with a pure integer program

(1) $$x_j \geq 0, \quad j = 1, \cdots, n,$$

(2) $$x_j \text{ integer}, \quad j = 1, \cdots, n,$$

(3) $$Ax = b,$$

(4) $$cx = z \text{(minimize)}.$$

When the problem is solved as a linear program, ignoring (2), some of the basic variables typically take on noninteger values. For any such basic variable, the updated row of the tableau is an equation of the form

(5) $$x_k + \sum_{j \in J_N} \alpha_j x_j = \beta,$$

where x_k is a basic variable and J_N is the set of nonbasic indices. The α_j and β are real numbers. The fractional cut of Gomory is the inequality

(6) $$\sum_{j \in J_N} \mathscr{F}(\alpha_j) x_j \geq \mathscr{F}(\beta),$$

where $\mathcal{F}(\alpha) = \alpha - \lfloor \alpha \rfloor$, for any real number α, and where $\lfloor \alpha \rfloor$ is the largest integer less than or equal to α. To derive (6), from (5) and $\alpha_j = \lfloor \alpha_j \rfloor + \mathcal{F}(\alpha_j)$ we see that

(7) $$\sum_{j \in J_N} \mathcal{F}(\alpha_j) x_j = \mathcal{F}(\beta) + \left(\lfloor \beta \rfloor - x_k - \sum_{j \in J_N} \lfloor \alpha_j \rfloor x_j \right).$$

By $x_j \geq 0, j \in J_N$, and $\mathcal{F}(\alpha_j) \geq 0$, we know that

(8) $$\sum_{j \in J_N} \mathcal{F}(\alpha_j) x_j \geq 0.$$

By x_k integer and x_j integer, $j \in N$, we have that

(9) $$\lfloor \beta \rfloor - x_k - \sum_{j \in J_N} \lfloor \alpha_j \rfloor x_j$$

is an integer. Hence, $\sum \mathcal{F}(\alpha_j) x_j$ differs from $\mathcal{F}(\beta)$ by an integer. But it cannot be $\mathcal{F}(\beta) - 1$ or smaller because of (8) and $\mathcal{F}(\beta) < 1$. Hence, (6) follows.

Let us review what was needed to prove (6). First, (5) was used. Then, $x_j \geq 0, j \in J_N$, was used to get (8). Finally, x_k and $x_j, j \in J_N$, integer was used to show that (9) was an integer. Note that $x_k \geq 0$ was not used. We shall return to that point.

Notice also that (6) is equivalent, as a linear inequality in conjunction with (5), to

(10) $$\lfloor \beta \rfloor \geq x_k + \sum_{j \in J_N} \lfloor \alpha_j \rfloor x_j$$

since, in fact, the slack variable to (6) is equal to the expression (9) and non-negativity of the slack is the same as (10). The fact that the slack will be integer when x_k and $x_j, j \in J_N$, are integer is an important point when cuts are generated from previous cuts. That is, the slack variables can also be treated as integer variables.

From (5), Gomory showed [3] how to generate a set of cuts. If we multiply (5) through by an integer h, then the same derivation shows that

(11) $$\sum_{j \in N} \mathcal{F}(h\alpha_j) x_j \geq \mathcal{F}(h\beta).$$

There are only a finite number of such inequalities which can be generated in this way; in fact, at most D where D is the least common denominator of the numbers $\alpha_j, j \in N$. To see this fact, for any integer h,

$$\mathcal{F}((h + D)\alpha_j) = \mathcal{F}(h\alpha_j + D\alpha_j) = \mathcal{F}(h\alpha_j)$$

because $D\alpha_j$ is an integer, and adding an integer to $h\alpha_j$ does not change the fractional part of it. Thus, the set of inequalities (11) will repeat with period D.

It is this periodic, or cyclic, repetition that will be returned to in the context of the cyclic group problem.

Using Cramer's rule, we see that D is either equal to the determinant of the basis B or is a divisor of that determinant.

Exercise 1. Consider the integer program

$$2x_1 + 3x_2 + 3x_3 - 2x_4 + 5x_5 = 2,$$
$$-2x_1 \qquad + x_3 \qquad - 3x_5 = -1,$$
$$x_1 + x_2 + x_3 + 2x_4 + 3x_5 = z(\min),$$
$$x_j \geq 0 \text{ and integer}.$$

Solving, as a linear program, gives

$$x_1 \qquad - \tfrac{1}{2}x_3 \qquad + 1\tfrac{1}{2}x_5 = \tfrac{1}{2},$$
$$x_2 + 1\tfrac{1}{3}x_3 - \tfrac{2}{3}x_4 + \tfrac{2}{3}x_5 = \tfrac{1}{3},$$
$$\tfrac{1}{6}x_3 + 2\tfrac{2}{3}x_4 + \tfrac{5}{6} = z - \tfrac{5}{6}.$$

Generate all of the fractional cuts from this tableau.

3. The cutting stock problem. The cutting stock problem arises in the paper industry where there are orders for rolls of paper of specified lengths and the manufacturer has large rolls of paper from which to cut the rolls ordered. The objective is to use as few of the large rolls as possible.

Let the large rolls have length l_0 and let $l_1, \cdots, l_m$ be the lengths of the rolls ordered with $b_1, \cdots, b_m$ orders for the corresponding lengths. Call $u_1, \cdots, u_m$ a *cutting pattern* if each u_i is a nonnegative integer such that

$$\sum_{i=1}^{m} u_i l_i \leq l_0.$$

Then, a large roll can be cut into u_1 rolls of length $l_1, \ldots,$ and u_m rolls of length l_m. The cutting stock problem can now be posed as follows:

$$\text{minimize} \sum_{j=1}^{n} y_j \text{ subject to}$$

$$y_j \geq 0 \text{ and integer},$$
$$Uy \geq b,$$

where each column of U is a cutting pattern, and U includes among its columns all possible cutting patterns. The variable y_j specifies the number of large rolls to be cut using the cutting pattern $u_{1j}, \cdots, u_{mj}$.

In performing the simplex method, if $\pi_1, \cdots, \pi_m$ are the dual variables for a primal feasible basis, the basis is optimum if

$$\sum_{i=1}^{m} \pi_i u_i \leq 1$$

for all cutting patterns $u_1, \cdots, u_m$. Thus, to determine if the basis is optimum, or if not to find an entering column, we need to solve the problem

(12) $$u_i \geq 0 \text{ and integer,}$$

(13) $$\sum_{i=1}^{m} l_i u_i \leq l_0,$$

(14) $$\sum_{i=1}^{m} \pi_i u_i = v(\text{maximize}).$$

If the maximum $v > 1$, then that cutting pattern given by the optimum $u_1, \cdots, u_m$ is the new entering column.

This column generation procedure avoids having to explicitly find U, which may be a matrix with thousands of columns. Each new entering column is generated by solving the problem (12)–(14). This problem is called the *knapsack problem*. The next section discusses that problem.

4. The knapsack problem. There is a very simple way to solve the knapsack problem (12)–(14) based on the recursion

$$v(\lambda) = \max_{i=1,\cdots,m} \{v(\lambda - l_i) + \pi_i\},$$

beginning with $v(\lambda) = 0$ if $\lambda \leq 0$. The recursion is performed for $\lambda = 1, 2, \cdots, l_0$. The Gilmore–Gomory work [2] is based on variations of this recursion. However, they observed an interesting property of the solutions. We illustrate with an example.

Example 1. Consider the knapsack problem:

$$u_j \geq 0 \text{ and integer};$$

$$3u_1 + 5u_2 + 10u_3 \leq 22;$$

$$6u_1 + 9u_2 + 19u_3 = v(\max).$$

Solving gives us Table 1. Notice that, from $\lambda = 9$ on, there is a cyclic repetition in the values of u_2 and u_3. Further, u_1 increases by 1 every time λ increases by 3, and v increases by 6 (the objective coefficient for u_1) every time λ increases by 3.

TABLE 1

λ	1	2	3	4	5	6	7	8	9	10	11	12	13	14	15	16	17	18	19	20	21	22
v	0	0	6	6	9	12	12	15	18	19	21	24	25	27	30	31	33	36	37	39	42	43
u_1			1	1		2	2	1	3		2	4	1	3	5	2	4	6	3	5	7	4
u_2					1			1			1			1			1			1		
u_3										1			1			1			1			1
s	1	2		1			1															

To see the general situation for the knapsack problem (12)–(14), assume that $\pi_1/l_1 \geq \pi_j/l_j$, $j = 2, \cdots, m$. Solving as a linear program gives the equation

$$\text{(15)} \qquad u_1 + \sum_{j=2}^{m} \frac{l_j}{l_1} u_j + \frac{1}{l_1} s = \frac{l_0}{l_1},$$

where s is the slack variable, $s = l_0 - \sum l_j u_j$. Now, eliminating u_1 from (14) and multiplying through by -1 gives

$$\text{(16)} \qquad z = -v + \frac{\pi_1}{l_1} l_0 = \sum_{j=2}^{m} \left(\pi_1 \frac{l_j}{l_1} - \pi_j \right) u_j + \frac{\pi_1}{l_1} s.$$

Let us now constrain u and s by

(17) $\qquad u_1$ integer, $\quad u_2, \cdots, u_m$, and $s \geq 0$ and integer.

THEOREM 1 [2]. *If l_0 is large enough ($l_0 \geq l_1 \times \max\{l_2, \cdots, l_m, 1\}$ will suffice), then the optimum solution to the knapsack problem for $l'_0 = l_0 + l_1$ is obtained from that for l_0 by changing only u_1 to $u_1 + 1$, and v increases by l_1. Thus, the solutions repeat with period l_1, and there are only l_1 different values of $u_2, \cdots, u_m, s$ needed in any optimum solution, once l_0 is large enough.*

The proof of Theorem 1 requires only noting that the solution to (15)–(17) is changed as stated in the theorem when l_0 is increased to $l_0 + l_1$. The reason is that since u_1 does not appear in (16) and is required only to be integer, the problem can be stated without u_1 by restating (15) as

$$\sum_{j=2}^{m} \frac{l_j}{l_1} u_j + \frac{1}{l_1} s \equiv \frac{l_0}{l_1} \text{ (modulo 1)},$$

that is, the left hand side differs from the right hand side by an integer (namely, u_1). Since u_j and s are integer variables, we can equivalently restate (15) as

$$\text{(18)} \qquad \sum_{j=2}^{m} \mathscr{F}\left(\frac{l_j}{l_1}\right) u_j + \mathscr{F}\left(\frac{1}{l_1}\right) s \equiv \mathscr{F}\left(\frac{l_0}{l_1}\right) \text{ (modulo 1)}.$$

Now, replacing l_0 by $l_0 + l_1$ does not change $\mathscr{F}(l_0/l_1)$ so the problem remains identical.

The sufficiency condition for l_0 being large enough depends on showing

$$\sum_{j=2}^{m} u_j + s \leq l_1.$$

For a proof, see Gomory [5, Thm. 2].

REFERENCES

[1] P. C. GILMORE AND R. E. GOMORY, *A linear programming approach to the cutting stock problem*, Operations Res., 9 (1961), pp. 849–859.

[2] ——, *The theory and computation of Knapsack functions*, Ibid., 14 (1966), pp. 1045–1074.
[3] R. E. GOMORY, *An algorithm for integer solutions to linear programs*, Recent Advances in Mathematical Programming, R. L. Graves and P. Wolfe, eds., McGraw-Hill, New York, 1963, pp. 269–302.
[4] ——, *On the relation between integer and non-integer solutions to linear programs*, Proc. Nat. Acad. of Sci U.S.A., 53 (1965), pp. 260–265.
[5] ——, *Faces of an integer polyhedron*, Ibid., 57 (1967), pp. 16–18.
[6] ——, *Some polyhedra related to combinatorial problems*, Linear Algebra and Appl., 2 (1969), pp. 451–558.
[7] R. E. GOMORY AND E. L. JOHNSON, *Some continuous functions related to corner polyhedra*, Math. Programming, 3 (1972), pp. 23–85 and pp. 359–389.

CHAPTER III

Finite Abelian Groups

1. Cyclic groups. Before proceeding, let us digress by discussing the nature of cyclic groups. It is known that every Abelian (which is the type we have here), finite cyclic group is equivalent to the integers taken modulo some particular integer. We will simply think of cyclic groups as being such objects.

Given some positive integer D, consider the integers $0, 1, \cdots, D - 1$ with addition $\hat{+}$ defined by $i \hat{+} j$ (modulo D). That is, if $i + j$ (in the usual sense) is greater than or equal to D, then subtract D from $i + j$ to determine $i \hat{+} j$. Otherwise, $i \hat{+} j = i + j$. In this way, $\hat{+}$ is a closed operation on $\{0, 1, \cdots, D - 1\}$ in that every sum is again a member of $\{0, 1, \cdots, D - 1\}$.

The set $0, 1, \cdots, D - 1$ with addition $\hat{+}$ is the *cyclic group* $\mathscr{C}_D$ of order D. It is a rather simple mathematical object but does have some interesting properties. First, define a *subgroup* of $\mathscr{C}_D$ to be a subset $\mathscr{C}$ of $\{0, 1, \cdots, D - 1\}$ which is closed with respect to the addition $\hat{+}$. It can be shown that there is exactly one subgroup of $\mathscr{C}_D$ for every positive integer n which divides D. For such an n, the subgroup is

$$\left\{0, \frac{D}{n}, 2\frac{D}{n}, \cdots, (n-1)\frac{D}{n}\right\}$$

and is the same as $\mathscr{C}_n$.

When we speak of there being only one subgroup, we are using the notion of isomorphism. Generally, two finite groups $\mathscr{G} = \{g_1, \cdots, g_n\}$ with addition $+$ and $\hat{\mathscr{G}} = \{\hat{g}_1, \cdots, \hat{g}_n\}$ with addition $\hat{+}$ are *isomorphic* if there exists a one to one mapping ϕ from $\mathscr{G}$ onto $\hat{\mathscr{G}}$ such that if $g = g_1 + g_2$ then $\phi(g) = \phi(g_1) \hat{+} \phi(g_2)$. In other words, ϕ maps not only from $\mathscr{G}$ onto $\hat{\mathscr{G}}$ but from the addition table of $\mathscr{G}$ onto that of $\hat{\mathscr{G}}$.

If a mapping ϕ is a many to one mapping from $\mathscr{G}$ onto $\hat{\mathscr{G}}$, but addition still is preserved by ϕ, then ϕ is a *homomorphism*. It can be shown that there is a homomorphism of $\mathscr{C}_D$ for each proper subgroup, and that subgroup is the image of a homomorphism.

For cyclic groups, every *morphism,* i.e. every isomorphism or homomorphism, can be determined by knowing $\phi(1)$. It must be true that $\phi(0) = 0$. There are D different morphisms corresponding to $\phi(1) = 0, \cdot, 2, \cdots, D - 1$. The morphism is an isomorphism if the order (the *order* is the number of times an element must be added to itself before the sum becomes 0) of $\phi(1)$ is D; otherwise, it is a homomorphism.

Exercise 1. For the cyclic group of order 6, determine all morphisms. For each homomorphism, determine the subgroup it maps onto, and for each such subgroup give an isomorphism onto the usual representation of that subgroup as a cyclic group.

2. The cyclic group problem. The cyclic group problem [2], [3] is to minimize

$$z = \sum_{j=1}^{n} c_j t_j$$

subject to

(1) $\qquad\qquad t_j \geq 0$ and integer for $j = 1, \cdots, n$

and

(2) $\qquad\qquad \sum_{j=1}^{n} g_j t_j \equiv g_0$ (modulo D),

where the $g_j, j = 1, \cdots, n$, are positive integers, $g_j < D$. The reason for calling it the cyclic group problem is that $\{g_1, \cdots, g_n\}$ can be considered as elements of the cyclic group $\mathscr{C}_D$, and addition in (2) can be considered to be the addition $\hat{+}$ of $\mathscr{C}_D$. If g_j is in $\mathscr{C}_D$ and t_j is a nonnegative integer, then $g_j t_j$ can be defined in the group $\mathscr{C}_D$ as adding g_j to itself t_j times:

$$g_j t_j = g_j \hat{+} \cdots \hat{+} g_j.$$

Thus, the cyclic group problem is: given some subset $\{g_1, \cdots, g_n\}$ (not necessarily a subgroup) of $\mathscr{C}_D$ and a distinguished element g_0 of $\mathscr{C}_D$, find the cheapest way, where g_j cost c_j for each time it is used, of adding up the elements in the subset so as to get g_0.

We note that if some g_j were an integer but not between 0 and $D - 1$, then D could be added or subtracted from it enough times to bring it between 0 and $D - 1$ without changing (2).

The right hand side g_0 of (2) need not be one of $\{g_1, \cdots, g_n\}$. The problem has a solution if and only if g_0 lies in the subgroup generated by $\{g_1, \cdots, g_n\}$; that is, the subgroup of all sums of the form in (2) where t satisfies (1). Let us assume that g_0 does lie in that subgroup.

If any $c_j < 0$, then the objective function z is unbounded. The reason is that for any solution t^0 to (1) and (2), t_j^0 can be changed to $t_j^0 + kD$, for any positive integer k, without effecting feasibility while decreasing z, whenever $c_j < 0$. Thus, we should assume that $c_j \geq 0$.

The cyclic group problem is obtained from an updated row of a linear programming tableau (Chapter II, (5)) by multiplying through by the least common denominator D of $\alpha_j, j \in J_N$, and β. Then, dropping nonnegativity of x_k but keeping the integrality requirement gives a cyclic group problem. The fractional cut is an inequality satisfied by every solution to that cyclic group problem. This fact is seen by recalling that, as was pointed out, nonnegativity of x_k was not used in proving validity of the fractional cut.

The set of fractional cuts generated in § 2 can be viewed as fractional cuts generated from equivalent or relaxed versions of the original cyclic group problem. For the cyclic group problem (1) and (2) also satisfies

$$\sum_{j=1}^{n} \phi(g_j) t_j \equiv \phi(g_0) \tag{3}$$

in the cyclic group which ϕ maps onto. Any cut for this problem is satisfied by all solutions to (1) and (2).

Exercise 2. For the cyclic group problem

$$t_1, t_2, t_3 \geq 0 \text{ and integer,}$$

$$1t_1 + 4t_2 + 5t_3 \equiv 1 \pmod{6}$$

give all of the cyclic group problems derived by morphisms and their corresponding fractional cuts.

For the knapsack problem in Chapter II, § 4, a cyclic group problem is obtained from Chapter II (15) by multiplying through by l_1:

$$\sum_{j=2}^{m} l_j u_j + s \equiv l_0 \pmod{l_1}.$$

This is the problem whose solution gives $u_2, \cdots, u_m, s$ in the periodic range of knapsack solutions.

Exercise 3. For the knapsack problem in Example 1 of Chapter II, give the corresponding cyclic group problem.

Cyclic group problems can be solved using a shortest path technique. A network is formed with nodes $0, 1, \cdots, D-1$ and arcs having length c_j are placed from node $g, g = 0, 1, \cdots, D-1$, to node $g \hat{+} g_j$. The solution is obtained by finding the shortest distance path from node 0 to node g_0. A Dijkstra type algorithm (see [4, § 19.3]) can be used for its solution.

Exercise 4. Set up and solve the shortest path problem derived from the cyclic group problem of Exercise 3.

Note that if the same element of $\mathscr{C}_D$ appears more than once in $\{g_1, \cdots, g_n\}$, as can happen in the reduction modulo D, then we may as well use only the cheapest (in terms of c_j) one.

3. Finite Abelian groups. Let $\mathscr{C}^1, \cdots, \mathscr{C}^K$ be finite cyclic groups of order $D_1, \cdots, D_K$, respectively. The *direct product* $\mathscr{G} = \mathscr{C}^1 \times \cdots \times \mathscr{C}^K$ is the set of all vectors $g = (g^1, \cdots, g^K)$, $g^k \in \mathscr{C}^k$, with addition $\hat{+}$ defined on $\mathscr{G}$ by

$$g \hat{+} h = (g^1, \cdots, g^K) \hat{+} (h^1, \cdots, h^K) = (g^1 \hat{+}_1 h^1, \cdots, g^K \hat{+}_K h^K).$$

Then, $\mathscr{G}$ is a finite, Abelian group because it satisfies, for all g, h, and k in $\mathscr{G}$,

$g \hat{+} h \in \mathscr{G}$ (closure),

$g \hat{+} (h \hat{+} k) = (g \hat{+} h) \hat{+} k$ (associativity),

$g \hat{+} \hat{0} = g$, for $\hat{0} = (0, \cdots, 0)$ (existence of zero),

$g \hat{+} (D_1 \hat{-} g_1, \cdots, D_K \hat{-} g_K) = \hat{0}$ (existence of a negative),

$g \hat{+} h = h \hat{+} g$ (commutativity).

It can be shown that any finite Abelian group is isomorphic to a direct product of finite Abelian cyclic groups. We can, thus, think of $\mathcal{G}$ as consisting of vectors of integers, and component k is taken modulo D_k.

Different direct products of groups may or may not give isomorphic groups. We leave an example as an exercise.

Exercise 5. Give an isomorphism from $\mathcal{C}_2 \times \mathcal{C}_3$ onto $\mathcal{C}_6$, but show that none exists for $\mathcal{C}_2 \times \mathcal{C}_2$ and $\mathcal{C}_4$.

A unique representation is obtained by the fact that any cyclic group $\mathcal{C}^D$ can be written as $\mathcal{C}^D = \mathcal{C}^1 \times \cdots \times \mathcal{C}^K$ where

$$D = p_1^{l_1} \times \cdots \times p_K^{l_K}, \quad \text{each } p_k \text{ is a prime and } l_k \geq 1,$$

and $\mathcal{C}^k$ is the cyclic group of order $p_k^{l_k}$, $k = 1, \cdots, K$. Then, any finite Abelian group can be written as the direct product of cyclic groups each of which is of order equal to some power of a prime.

Example 1. The Abelian groups up to order 11 are $\mathcal{C}_1$, $\mathcal{C}_2$, $\mathcal{C}_3$, $\mathcal{C}_4$ and $\mathcal{C}_2 \times \mathcal{C}_2$, $\mathcal{C}_5$, $\mathcal{C}_2 \times \mathcal{C}_3$, $\mathcal{C}_7$, $\mathcal{C}_8$ and $\mathcal{C}_4 \times \mathcal{C}_2$ and $\mathcal{C}_2 \times \mathcal{C}_2 \times \mathcal{C}_2$, $\mathcal{C}_9$ and $\mathcal{C}_3 \times \mathcal{C}_3$, $\mathcal{C}_2 \times \mathcal{C}_5$, $\mathcal{C}_{11}$.

All of the subgroups of a finite Abelian group $\mathcal{G} = \mathcal{C}^1 \times \cdots \times \mathcal{C}^K$ can be formed by taking all of the direct products $\mathcal{D}^1 \times \cdots \times \mathcal{D}^K$ where each $\mathcal{D}^k$ is a subgroup of the cyclic group $\mathcal{C}^K$. All other subgroups are isomorphic to one of these.

Homomorphisms and subgroups are closely related in that every homomorphism ϕ maps onto a subgroup $\mathcal{H}$ of $\mathcal{G}$. There is another relation: the elements $g \in \mathcal{G}$ which ϕ maps onto 0 form a subgroup $\mathcal{G}_0$, called the *kernel of* ϕ. The subgroup $\mathcal{H}$ which ϕ maps onto is sometimes called the *factor group*

$$\mathcal{H} = \mathcal{G}/\mathcal{G}_0.$$

While knowing the kernal $\mathcal{G}_0$ of ϕ does not determine ϕ, it does determine the factor group $\mathcal{H} = \mathcal{H}/\mathcal{G}_0$.

Example 2. Consider the group $\mathcal{C}_2 \times \mathcal{C}_4$ with ϕ:

$$\mathcal{G} = \left\{ \begin{pmatrix} 0 \\ 0 \end{pmatrix}, \begin{pmatrix} 1 \\ 0 \end{pmatrix}, \begin{pmatrix} 0 \\ 1 \end{pmatrix}, \begin{pmatrix} 1 \\ 1 \end{pmatrix}, \begin{pmatrix} 0 \\ 2 \end{pmatrix}, \begin{pmatrix} 1 \\ 2 \end{pmatrix}, \begin{pmatrix} 0 \\ 3 \end{pmatrix}, \begin{pmatrix} 1 \\ 3 \end{pmatrix} \right\}$$

$$\phi \quad \downarrow \quad \downarrow \quad \downarrow \quad \downarrow \quad \downarrow \quad \downarrow \quad \downarrow \quad \downarrow$$

$$\begin{pmatrix} 0 \\ 0 \end{pmatrix} \begin{pmatrix} 1 \\ 0 \end{pmatrix} \begin{pmatrix} 0 \\ 1 \end{pmatrix} \begin{pmatrix} 1 \\ 1 \end{pmatrix} \begin{pmatrix} 0 \\ 0 \end{pmatrix} \begin{pmatrix} 1 \\ 0 \end{pmatrix} \begin{pmatrix} 0 \\ 1 \end{pmatrix} \begin{pmatrix} 1 \\ 1 \end{pmatrix}.$$

The kernel $\mathcal{G}_0$ and factor group $\mathcal{F}$ are given below:

$$\mathcal{H} = \mathcal{C}_2 \times \mathcal{C}_2 = \mathcal{G} \Big/ \left\{ \begin{pmatrix} 0 \\ 0 \end{pmatrix}, \begin{pmatrix} 0 \\ 2 \end{pmatrix} \right\}.$$

If the kernel is changed to

$$\left\{ \begin{pmatrix} 0 \\ 0 \end{pmatrix}, \begin{pmatrix} 1 \\ 0 \end{pmatrix} \right\},$$

which is isomorphic to the above kernel, then the factor group is changed from $\mathscr{C}_2 \times \mathscr{C}_2$ to $\mathscr{C}_4$. Thus, the actual subgroup forming the kernel is needed in order to know the factor group.

REFERENCES

[1] L. FUCHS, *Abelian Groups*, Pergamon Press, New York, 1960.
[2] R. E. GOMORY, *On the relation between integer and non-integer solutions to linear programs*, Proc. Nat. Acad. of Sci. U.S.A., 53 (1965), pp. 260–265.
[3] ———, *Some polyhedra related to combinatorial problems*, Linear Algebra and Appl., 2 (1969), pp. 451–558.
[4] T. C. HU, *Integer Programming and Network Flows*, Addison-Wesley, Reading, MA, 1969.

CHAPTER IV

Gomory's Corner Polyhedra

1. The group problem. Given a finite Abelian group $\mathcal{G}$, an element $g_0 \in \mathcal{G}$, and a subset $\mathcal{S}$ of $\mathcal{G}$, the group problem is

(1) $\qquad t(g) \geq 0$ and integer, $\qquad g \in \mathcal{S}$,

(2) $\qquad \sum_{g \in \mathcal{S}} gt(g) = g_0,$

(3) $\qquad \text{minimize} \sum_{g \in \mathcal{S}} c(g)t(g),$

where $c(g)$, $g \in \mathcal{S}$, are real numbers.

The notation here is different from that in Chapter III, § 2 in that t here is simply indexed by $g \in \mathcal{S}$ rather than numbering the $g \in \mathcal{S}$ and indexing t by the index of g. Here, the equation (2) is with respect to addition for the group $\mathcal{G}$. As in Chapter III, § 2, $gt(g)$ is the group element obtained by adding g to itself $t(g)$ times.

As in Chapter II, § 2, we assume g_0 is in the subgroup generated by $\mathcal{S}$ so that there exists a solution t to (1) and (2). Also, $c(g) \geq 0$ can be assumed since otherwise the objective (3) is unbounded. Then, there is an optimum t satisfying $t(g) \leq D$, all $g \in \mathcal{L}$, where $D = |\mathcal{G}|$ is the order of $\mathcal{G}$.

The group problem is a relaxation of the integer programming problem of minimizing a linear objective in x subject to $Ax = b$, $x \geq 0$ and integer. Assume that A and b have integer elements. Solving the associated linear program gives a partition of A and x into

(4) $\qquad\qquad\qquad Bx_B + Nx_N = b.$

If we impose the integrality requirement on (x_B, x_N) but drop $x_B \geq 0$, the constraints become

(5) $\qquad\qquad Nx_N \equiv b(\text{modulo } B), \qquad x_N \geq 0$ and integer.

Here, modulo B means that Nx_N and b are considered to be group elements in the factor group

$$\mathcal{G} = Z^m/Z(B),$$

where $Z^m = \{(i_1, \cdots, i_m)|i_j \text{ integer}\}$ and

$$Z(B) = \{Bx_B|x_B \text{ integer}\}.$$

Since B is a basis of $\mathcal{R}^m$, $\mathcal{G}$ is a finite group. What group it is, as a direct product of cyclic groups, can be determined by reducing it using unimodular transformation U and V to a diagonal matrix M:

$$M = UBV.$$

An equivalent way of looking at this relaxation is to update the linear programming optimal tableau to obtain

$$\bar{N}x_N \equiv \bar{b} \text{ (modulo 1)}, \qquad x_N \geq 0 \text{ and integer}.$$

Since A has integer coefficients, the entries in $\bar{N}$ will all be of the form p/D where p is an integer and $D = \det(B)$, by Cramer's rule. Thus, we can consider the group to be

$$\bar{\mathcal{G}} = \left\{ \left(\frac{p_1}{D}, \ldots, \frac{p_m}{D}\right) \middle| 0 \leq p_i < D, p_i \text{ integer} \right\}$$

with addition taken modulo 1; i.e., all components are reduced to their fractional parts. This group is of order D^m while $\mathcal{G} = Z^m/Z(B)$ is only of order D. However, the columns of $\bar{N}$ will generate a subgroup of order D. The much larger order of $\bar{\mathcal{G}}$ may not be of consequence if one is using a method of solving the group problem which only looks at group elements g generated.

If two or more columns of N map onto the same group element, then one with least cost can be chosen and the rest set to zero since there is always an optimum group solution with them at value zero. If any column of N maps onto the zero of the group, then its variable can also be set to zero.

Example 1. Consider the integer program:

$$x_1, x_2, x_3 \geq 0 \text{ and integer};$$
$$6x_1 + 10x_2 + x_3 \geq 20;$$
$$4x_1 + 3x_2 + 9x_3 \geq 13;$$
$$\text{minimize } z = 3x_1 + 6x_2 + 4x_3.$$

An optimum linear programming tableau is:

$$x_1 \quad + 1\tfrac{2}{3}x_2 + \tfrac{1}{6}x_3 - \tfrac{1}{6}s_1 = 3\tfrac{1}{3};$$
$$s_2 + 3\tfrac{2}{3}x_2 - 8\tfrac{1}{3}x_3 - \tfrac{2}{3}s_1 = \tfrac{1}{3};$$
$$x_2 + 3\tfrac{1}{2}x_3 + \tfrac{1}{2}s_1 = z - 10.$$

A representation of the group problem is:

$$\tfrac{2}{3}x_2 + \tfrac{1}{6}x_3 + \tfrac{5}{6}s_1 \equiv \tfrac{1}{3} \text{ (modulo 1)},$$
$$\tfrac{2}{3}x_2 + \tfrac{2}{3}x_3 + \tfrac{1}{3}s_1 \equiv \tfrac{1}{3} \text{ (modulo 1)}.$$

The columns which can actually be generated by nonnegative x_2, x_3, s_1 are:

$$\begin{pmatrix}0\\0\end{pmatrix}, \begin{pmatrix}\tfrac{1}{6}\\\tfrac{2}{3}\end{pmatrix}, \begin{pmatrix}\tfrac{1}{3}\\\tfrac{1}{3}\end{pmatrix}, \begin{pmatrix}\tfrac{1}{2}\\0\end{pmatrix}, \begin{pmatrix}\tfrac{2}{3}\\\tfrac{2}{3}\end{pmatrix}, \begin{pmatrix}\tfrac{5}{6}\\\tfrac{1}{3}\end{pmatrix}.$$

Thus subgroup is isomorphic to $\mathscr{C}_6$ and, in fact, the constraint of the group problem is the same as

$$4x_2 + 1x_3 + 5\, s_1 \equiv 2 \text{ (modulo 6)}.$$

2. The asymptotic theorem. Gomory's theorem [1] to be discussed here can be viewed in more than one way. An essential point is: When does the solution to the group problem (5) satisfy the integer problem? The nonbasic variables x_N are given by (5), and the basic variables by

(6) $$x_B = B^{-1}(b - Nx_N).$$

Because of (5), x_B will be integer, but the nonnegativity requirement in the integer program was dropped so that x_B may not satisfy $x_B \geq 0$.

Consider the cone $\mathscr{K}^B$ of those $y \in \mathscr{R}^m$ such that $B^{-1}y \geq 0$. Clearly, $b \in \mathscr{K}^B$. Now if b is far enough in the interior of the cone, then

$$B^{-1}(b - Nx_N) \geq 0$$

will hold for all x_N satisfying

$$\sum_{j \in J_N} x_j \leq D - 1,$$

when J_N is the index set of nonbasic variables x_N. Since every optimum solution to the group problem satisfies the above inequality, where $D = \det(B)$, the x_B given by (6) will satisfy $x_B \geq 0$ provided b is at least the distance $(D-1)l$ from the boundary of $\mathscr{K}^B$, where

$$l = \max_{j \in J_N} \|N^j\|.$$

Thus, for all b far enough interior to $\mathscr{K}^B$, there are only D different values of x_N needed in optimum integer solutions; these values of x_N are obtained by solving (5); x_N is the same for $b + B^i$ as for b; and the values of x_B are given by (6).

Another way to view the theorem is that if we have a nondegenerate linear programming solution $x_B = B^{-1}b > 0$, then changing b to λb, $\lambda \geq 1$ and integer, will eventually bring λb, for large enough λ, into the periodic range where an optimum integer programming solution has x_N the same for $\lambda' = \lambda + D$ as for λ, and x_B increases by $D(B^{-1}b)$ for such a change in λ.

In this way, the result very much resembles the periodic behavior of the knapsack solutions. Also, it motivates the notion of asymptotic behavior of integer programs; if we start with a b giving a nondegenerate linear programming solution, then for the integer programs given by the set of right hand sides λb, $\lambda \geq 1$ and integer, as λ becomes large the group problem (5) with x_B given by (6) will solve the integer program.

All too often the focus is on the negative aspect of the asymptotic theorem, i.e., the fact that $x_B \geq 0$ may not always be satisfied. We even see the group problem referred to as the "asymptotic relaxation." The only sense in which asymptotic is used validly is the sense of describing behavior of integer programming solutions for large (in some sense) right hand sides b. There are few descriptive theorems in integer programming, and this theorem is one of those few. It can be compared to rounding, i.e., changing basic variables at fractional values to their nearest integer values. However, rounding always requires changing some of the nonbasic variables, even if only nonbasic slack variables. Here, the focus is on "rounding" the nonbasic variables first, but in such a way that the basic variables turn out to have integer values. How far the nonbasic variables have to be "rounded" is bounded by

$$\sum_{j \in J_N} x_j \leq D - 1,$$

where $D = \det(B)$. The values of x_B given by (6) may, of course, not be the nearest integers to their fractional values $B^{-1}b$; in fact, they can be quite far from the nearest integers.

Exercise 1. In Example 1, consider the group problems arising by letting the right hand side be

$$\lambda \begin{pmatrix} 20 \\ 13 \end{pmatrix}, \quad \lambda \geq 1 \text{ and integer.}$$

When is λ large enough for the periodic behavior to be true? Give the form of solutions to the integer program in the periodic range.

3. Corner polyhedra. For the group problem with constraints (1) and (2), define *Gomory's corner polyhedron* to be the convex hull of all vectors $(t(g), g \in \mathscr{G})$ satisfying (1) and (2). Let us denote this polyhedron by $\mathscr{P}(\mathscr{G}, \mathscr{S}, g^0)$.

We have assumed that there is some vector t^0 satisfying (1) and (2). For such a t^0, adding any positive multiple of D to any component of t^0 gives another solution to (1) and (2). Hence, if $x \in \mathscr{P}(\mathscr{G}, \mathscr{S}, g^0)$ and $y \geq 0$, then $x + y \in \mathscr{P}(\mathscr{G}, \mathscr{S}, g^0)$. Denoting the nonnegative orthant by $\mathscr{R}^n_+$, we have

(7) $$\mathscr{R}^n_+ \supseteq \mathscr{P}(\mathscr{G}, \mathscr{S}, g^0) = \mathscr{P}(\mathscr{G}, \mathscr{S}, g^0) + \mathscr{R}^n_+.$$

Next, let us establish that $\mathscr{P}(\mathscr{G}, \mathscr{S}, g^0)$ is, indeed, a polyhedron. A *polyhedron* is defined to be a convex set $\mathscr{P}$ of the form

$$\mathscr{P} = \text{conv}(V) + \text{cone}(R)$$

where V and R are finite subsets of $\mathscr{R}^n$ and

$$\text{conv}(V) = \left\{ \sum_{v \in V} v \lambda(v) \mid \lambda(v) \geq 0, \sum_{v \in V} \lambda(v) = 1 \right\},$$

$$\text{cone}(R) = \left\{ \sum_{r \in R} r \alpha(r) \mid \alpha(r) \geq 0 \right\}.$$

So far, $\mathscr{P}(\mathscr{G}, \mathscr{S}, g^0)$ has been defined as the convex hull of an infinite set of points; namely all t satisfying (1) and (2).

Let V be the (finite) set of all t satisfying (1), (2), and $0 \leq t_j < D$, and let R be the set of unit vectors $e^j = (0, \cdots, 0, 1, 0, \cdots, 0), j = 1, \cdots, n$, with a 1 in the jth place. Let $\mathcal{P} = \text{conv}(V) + \text{cone}(R)$. For any t satisfying (1) and (2), any coordinate j with $t_j \geq D$ can be reduced by subtracting a positive integral multiple of D without violating (1) or (2). Hence, $t \in \mathcal{P}$, and $\mathcal{P}(\mathcal{G}, \mathcal{S}, g^0) \subseteq \mathcal{P}$. The reverse inequality also holds because for any t in V, we can add nonnegative multiples of D to any coordinate without leaving $\mathcal{P}(\mathcal{G}, \mathcal{S}, g^0)$. Hence, $\mathcal{P}(\mathcal{G}, \mathcal{S}, g^0)$ is a polyhedron.

The polyhedron $\mathcal{P}(\mathcal{G}, \mathcal{S}, g^0)$ is in the nonnegative orthant $\mathcal{R}_+^n$ and is full dimensional. Consider the set of inequalities in

(8) $$\sum_{g \in \mathcal{S}} \pi(g) t(g) \geq \pi_0,$$

which are satisfied by all $t \in \mathcal{P}(\mathcal{G}, \mathcal{S}, g^0)$. For the same reason that $c(g) \geq 0$ was assumed in the previous section, every $\pi(g) \geq 0$ can be assumed. If not, then $\sum \pi(g) t(g)$ would be arbitrarily small for some $t \in \mathcal{P}(\mathcal{G}, \mathcal{S}, g^0)$ so the inequality (8) could not hold for any number π_0. Thus $\pi(g) \geq 0$ must hold. Then, $\pi_0 \geq 0$ can be assumed since $\sum \pi(g) t(g) \geq 0$ must be true just by $t(g) \geq 0$.

Clearly, $t(g) \geq 0$ is an inequality satisfied by all $t \in \mathcal{P}(\mathcal{G}, \mathcal{S}, g^0)$. Any other inequality (8) with $\pi_0 = 0$ can be written as a nonnegative linear combination of the inequalities $t_j \geq 0$, $j = 1, \cdots, n$. Hence, we focus on inequalities with $\pi_0 > 0$ and may scale π so that $\pi_0 = 1$.

Define a *valid inequality* for $\mathcal{P}(\mathcal{G}, \mathcal{S}, g^0)$ to be a vector $(\pi(g), g \in \mathcal{S})$ such that

(9) $$\sum_{g \in \mathcal{S}} \pi(g) t(g) \geq 1$$

for all $t \in \mathcal{P}(\mathcal{G}, \mathcal{S}, g^0)$. We have already shown that $\pi(g) \geq 0$ must hold for any valid inequality. Since (9) must hold for all $t \in \mathcal{P}(\mathcal{G}, \mathcal{S}, g^0)$, it must hold in particular for $t \in V$; i.e. t satisfying (1), (2), and $0 \leq t(g) < D$. We claim, in fact, that the set of valid inequalities for $\mathcal{P}(\mathcal{G}, \mathcal{S}, g^0)$ is the same as the set of $(\pi(g), g \in \mathcal{S})$ satisfying

$$\pi(g) \geq 0, \quad g \in \mathcal{S}$$

and

$$\sum_{g \in \mathcal{S}} \pi(g) t(g) \geq 1, \quad t \in V.$$

That is, any $(\pi(g), g \in \mathcal{S})$ satisfying nonnegativity and (9) for $t \in V$ will satisfy (9) for all $t \in \mathcal{P}(\mathcal{G}, \mathcal{S}, g^0)$. This fact can easily be seen by the representation

$$\mathcal{P}(\mathcal{G}, \mathcal{S}, g^0) = \text{conv}(V) + \text{cone}(R)$$

already established.

Gomory gave a sharper result in [3, § 1.B]. Define a t satisfying (1) and (2) to be *irreducible* if every integral t', $0 \leq t'(g) \leq t(g)$, $g \in \mathcal{S}$, gives a different group element $h(t') \in \mathcal{G}$ defined by

$$h(t') = \sum_{g \in \mathscr{G}} g t'(g).$$

Since there are

$$\prod_{g \in \mathscr{G}} (1 + t(g))$$

different such vectors t', and $|\mathscr{G}|$ different group elements, we have

$$\prod_{g \in \mathscr{G}} (1 + t(g)) \leq |\mathscr{G}|$$

for any irreducible t.

The definition of irreducible used in [2] was weaker than in [3], but the result was established there that $\pi(g) \geq 0$, $g \in \mathscr{G}$, and (9) for irreducible t defines the set of valid inequalities.

REFERENCES

[1] R. E. GOMORY, *On the relation between integer and non-integer solutions to linear programs*, Proc. Nat. Acad. of Sci. U.S.A., 53 (1965), pp. 260–265.

[2] ———, *Faces of an integer polyhedron*, Ibid., 57 (1967), pp. 16–18.

[3] ———, *Some polyhedra related to combinatorial problems*, Linear Algebra and Appl., 2 (1969), pp. 451–558.

CHAPTER V

Blocking Polyhedra and Master Group Problems

1. Blocking pairs of polyhedra. We now turn to Fulkerson's [3] framework of blocking pairs of polyhedra. Given a polyhedron $\mathcal{P}$ contained in $\mathcal{R}_+^n$, define the *blocking polyhedron* $\mathcal{B}(\mathcal{P})$ of $\mathcal{P}$ to be

$$\mathcal{B}(\mathcal{P}) = \{x^* \geq 0 | x^* \cdot x \geq 1 \text{ for all } x \in \mathcal{P}\}.$$

In order for

$$\mathcal{B}(\mathcal{B}(\mathcal{P})) = \mathcal{P}$$

to hold, it is necessary and sufficient [1] that

$$\mathcal{P} = \mathcal{P} + \mathcal{R}_+^n.$$

For generalizations and other discussion of this closure, see Araoz [1], Johnson [7], and Tind [9].

If the polyhedron $\mathcal{P}$ is given by

$$\mathcal{P} = \text{conv}(V) + \mathcal{R}_+^n,$$

then it can be shown that

$$\mathcal{B}(\mathcal{P}) = \{x^* \geq 0 | x^* \cdot v \geq 1, v \in V\}.$$

A minimal set V so defining $\mathcal{P}$ are the *vertices* of $\mathcal{P}$, and by the above result are the facets of $\mathcal{B}(\mathcal{P})$.

The blocking polyhedron $\mathcal{B}(\mathcal{P})$ has a similar representation:

$$\mathcal{B}(\mathcal{P}) = \text{conv}(V^*) + \mathcal{R}_+^n.$$

For a minimal such V^*, the $v^* \in V^*$ are the vertices of $\mathcal{B}(\mathcal{P})$ and the facets of $\mathcal{P}$, i.e.,

$$\mathcal{P} = \{x \geq 0 | x \cdot v^* \geq 1, v^* \in V^*\}.$$

When $\mathcal{P}$ and $\mathcal{B}(\mathcal{P})$ are so defined, then they are called a *blocking pair of polyhedra* and the matrices V and V^* are called a *blocking pair of matrices* [3]. There is a complete symmetry in both $\mathcal{P}$, $\mathcal{B}(\mathcal{P})$ and V, V^*.

Gomory's result [4] is concerned with the blocker $\mathcal{B}(\mathcal{P})$ of the polyhedron $\mathcal{P} = \mathcal{P}(\mathcal{G}, \mathcal{S}, g_0)$. He shows that the vertices of $\mathcal{P}$ are all irreducible, so that the set V of all irreducible solutions includes all vertices and some more vectors. In any case

$$\mathcal{B}(\mathcal{P}) = \{\pi \geq 0 | \pi \cdot t \geq 1 \text{ for all irreducible } t\},$$

and Gomory calls any $\pi \in \mathcal{B}(\mathcal{P})$ a valid inequality [5], [6].

Whereas the blocking theory is symmetric in $\mathcal{P}$ and $\mathcal{B}(\mathcal{P})$, Gomory's work is not symmetric. For example, the blocking theory includes a max min equality and a length width inequality [3] which are completely symmetric. It is interesting to attempt to interpret those results for the group problem.

Exercise 1*. Gomory shows that if every group element has order 2 or if every group element has order 3, then all irreducible solutions are vertices. For groups up to order 11, there is only one irreducible vector which is not a vertex. Try to establish other conditions when all irreducible solutions are vertices. Try to characterize the vertices exactly. See [10] for an attempt at characterization which proved to be incorrect.

2. Master group problems. Of particular importance is the case $\mathcal{S} = \mathcal{G}_+$, where $\mathcal{G}_+$ denotes the nonzero elements of $\mathcal{G}$. We denote the polyhedron in this case by $\mathcal{P}(\mathcal{G}, g_0)$ and call it the master polyhedron for $\mathcal{G}, g_0$. The subset $\mathcal{S}$ of $\mathcal{G}$ can be assumed to be a subset of $\mathcal{G}_+$ since if $\hat{0} \in \mathcal{S}$, $t(\hat{0}) = 0$ can always be assumed in any answer.

The polyhedron $\mathcal{P}(\mathcal{G}, \mathcal{S}, g_0)$ for any $\mathcal{S} \subseteq \mathcal{G}$ is obtained from $\mathcal{P}(\mathcal{G}, g_0)$ by setting $t(g) = 0$ for all $g \notin \mathcal{S}$. But $t(g) \geq 0$ is a face of $\mathcal{P}_\mathcal{G}(g_0)$ so $\mathcal{P}(\mathcal{G}, \mathcal{S}, g_0)$ is an intersection of $\mathcal{P}(\mathcal{G}, g_0)$ with a face of itself. This general polyhedral situation implies: (i) the vertices of $\mathcal{P}(\mathcal{G}, \mathcal{S}, g_0)$ are the vertices of $\mathcal{P}(\mathcal{G}, g_0)$ for which $t(g) = 0$, $g \notin \mathcal{S}$; (ii) the facets of $\mathcal{P}(\mathcal{G}, \mathcal{S}, g_0)$ are among the inequalities

$$\sum_{g \in \mathcal{S}} \pi(g) t(g) \geq 1$$

one gets from restricting a facet π for $\mathcal{P}(\mathcal{G}, g_0)$ to $\mathcal{S}$. Gomory [6] lists the incidence matrix of vertices versus facets for master problems up to order 11. From an incidence matrix of this type one can determine exactly which inequalities are facets. The facets are those inequalities for which the vertices satisfying the inequality with equality are not a subset of those for some other inequality.

Example 1. For $\mathcal{G} = \mathcal{C}_6$ and $g_0 = 3$, let $\mathcal{S} = \{1, 2, 5\}$. Gomory [6, Appendix 5] lists seven vertices, of which four involve only 1, 2, and 5: $t_1 = 3$, $t_1 = t_2 = 1$, $t_2 = 2$ and $t_5 = 1$, $t_5 = 3$. These vertices are 1, 2, 5, and 7 in the list, so we are interested in rows 1, 2, 5, and 7 in the incidence matrix as shown in Table 1. Because 3 and 4 are the deleted group elements, rows 3 and 4 and columns 7 and 8 are deleted. Columns 2, 3, 5, 6, and 9 are maximal columns, so the facets of $\mathcal{P}(\mathcal{C}_6, \{1, 2, 5\}, 3)$ are

$$2t_1 + t_2 + t_5 \geq 3,$$
$$t_1 + 2t_2 + t_5 \geq 3,$$
$$t_1 \geq 0,$$
$$t_2 \geq 0,$$
$$t_5 \geq 0.$$

TABLE 1

		Face									
		1	2	3	4	5	6	7	8	9
unit vectors	1	0	0	0	0	0	1	1	1	1
	2	1	0	0	0	1	0	1	1	1
	3	0	0	0	0	1	1	0	1	1
	4	1	0	0	0	1	1	1	0	1
vertex	5	0	0	0	0	1	1	1	1	0
	1	0	0	1	1	0	1	1	1	1
	2	1	1	1	1	0	0	1	1	1
	5	1	1	0	0	1	0	1	1	0
	7	0	1	1	0	1	1	1	1	0

3. Facets of master group problems.

THEOREM 1 [6, Thm. 18]. *The facets*

$$\sum_{g \in \mathcal{G}_+} \pi(g) t(g) \geq 1$$

of $\mathcal{P}(\mathcal{G}, g_0)$ *are the vertices of the polytope defined by*

(1) $\qquad 0 \leq \pi(g), \quad \text{all } g \in \mathcal{G}_+,$

(2) $\qquad \pi(g \dotplus h) \leq \pi(g) + \pi(h), \quad \text{all } g, h, \text{ and } g + h \in \mathcal{G}_+,$

(3) $\qquad \pi(g_0) = \pi(g) + \pi(g_0 \dotminus g), \quad \text{all } g \in \mathcal{G}_+,$

(4) $\qquad \pi(g_0) = 1.$

The proof of this theorem will be included in that of a more general theorem to be proven later.

Let us compare this result with the one that says the facets are vertices of

$$\pi(g) \geq 0, \quad g \in \mathcal{G}_+,$$

$$\sum_{g \in \mathcal{G}_+} \pi(g) t(g) \geq 1, \quad \text{all irreducible } t.$$

Example 2. The problem with constraints

$$t_j \geq 0, \quad j = 1, 2, 3,$$

and

$$t_1 + 2t_2 + 3t_3 \equiv 3 \pmod{4}$$

has three irreducible solutions: (3, 0, 0), (1, 1, 0), and (0, 0, 1). The corresponding inequalities are

$$\pi_1 \geq 0, \quad \pi_2 \geq 0, \quad \pi_3 \geq 0, \text{ and}$$

$$3\pi_1 \geq 1,$$

$$\pi_1 + \pi_2 \geq 1,$$

$$\pi_3 \geq 1.$$

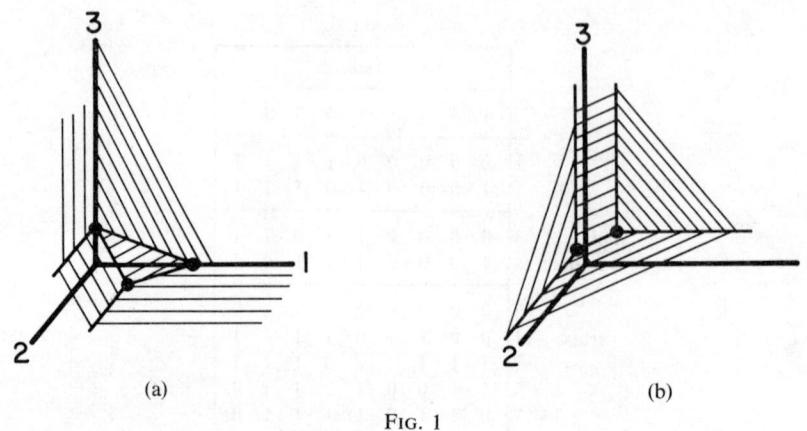

Fig. 1

In Fig. 1 (a), Gomory's corner polyhedron is shown, and Fig. 1 (b) shows the polyhedron of valid inequalities. These two polyhedra are blocking polyhedra.

Looking at Theorem 1, the inequalities there are

$$\pi_1 \geqq 0, \quad \pi_2 \geqq 0, \quad \pi_3 \geqq 0,$$
$$\pi_2 \leqq 2\pi_1,$$
$$\pi_3 = \pi_1 + \pi_2,$$
$$\pi_3 = 1.$$

Figure 2 illustrates this polyhedron.

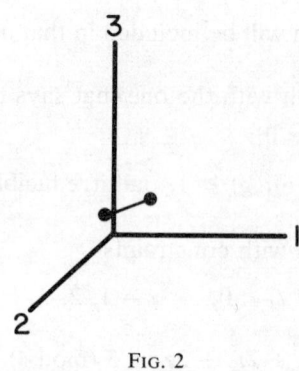

Fig. 2

Theorem 1 describes the facets of $\mathcal{P}(\mathcal{G}, g_0)$ as the vertices of a (bounded) polytope; i.e., as the vertices of their convex hull. By comparison, the blocking description describes the facets of $\mathcal{P}(\mathcal{G}, g_0)$ as the vertices of an unbounded polyhedron whose recession cone is, in fact, the nonnegative orthant $\mathcal{R}_+^n$.

4. Automorphisms and homomorphisms.

THEOREM 2 [6, Thm. 14]. *If ϕ is an automorphism of $\mathcal{G}$ onto itself and if π' is a face of $\mathcal{P}(\mathcal{G}, \phi(g_0))$, then π is a face of $\mathcal{P}(\mathcal{G}, g_0)$ where*

$$\pi(g) = \pi'(\phi(g)), \quad g \in \mathcal{G}_+.$$

Proof. The proof is evident from Theorem 1. The system of inequalities whose vertices are facets of $\mathcal{P}(\mathcal{G}, g_0)$ is given by (1)–(4). The system

$$0 \leq \pi(\phi(g))$$
$$\pi(\phi(g \hat{+} h)) \leq \pi(\phi(g)) + \pi(\phi(h)),$$
$$\pi(\phi(g_0)) = \pi(\phi(g)) + \pi(\phi(g_0 \hat{-} g)),$$
$$\pi(\phi(g_0)) = 1,$$

is exactly the same system except for the reordering of the indices g given by ϕ.

In his tables, Gomory only gives one listing of facets and vertices (a similar theorem holds for vertices) for a given $\mathcal{G}$ and the various g_0 for which there exist automorphisms mapping the g_0 onto each other. The fact that the relation defined by $g_1 \mathcal{R} g_2$ if there exists an automorphism ϕ of $\mathcal{G}$ with $\phi(g_1) = g_2$ partitions $\mathcal{G}$ into equivalence classes is proven by the fact that the automorphisms of a group $\mathcal{G}$ form a group themselves.

THEOREM 3 [6, Thm. 9]. *If ψ is a homomorphism of $\mathcal{G}$ onto $\mathcal{H}$ with kernel $\mathcal{K}$ having $g_0 \notin \mathcal{K}$, and if π' is a facet of $\mathcal{P}(\mathcal{H}, \psi(g_0))$, then π is a facet of $\mathcal{P}(\mathcal{G}, g_0)$ where $\pi(g) = \pi'(\psi(g))$. (We take $\pi'(\hat{0}) = 0$, so $\pi(g) = 0, g \in \mathcal{K}$.)*

Proof. Form the matrix S with a row for each $g \in \mathcal{G}$ (including $\hat{0}$) and a column for each of the inequalities $0 \leq \pi(g) + \pi(h) - \pi(g + h)$, $0 \leq \pi(g)$, and $1 \leq \pi(g_0)$. The automorphism ψ maps S onto a matrix $\psi(S)$ obtained by adding all of the rows which map onto the same element $h \in \mathcal{H}$ and removing duplicate columns. Now, $\psi(S)$ represents the same inequalities $0 \leq \pi'(g') + \pi'(h') - \pi'(g' + h'), 0 \leq \pi'(g')$, and $1 \leq \pi'(g_0')$ in $\mathcal{H}$. A facet π' of $\mathcal{P}(\mathcal{H}, \psi(g_0))$ is a solution for which equality holds corresponding to a basis of $\psi(S)$.

The π as defined in the statement of the theorem is easily seen to satisfy the inequalities (1)–(4). It remains to show that the columns of S where equality holds include a basis of S. Suppose not. Then, restricting ourselves to those columns, there is a linear combination $\lambda(g), g \in \mathcal{G}$, of the rows of S giving the zero row in those columns. For $g \in \mathcal{K}$, $\pi(g) = \pi'(\psi(g)) = 0$, so $\lambda(g) = 0$ for all $g \in \mathcal{K}$. Since $\pi(g_0) = \pi'(\psi(g_0)) = 1$, $\pi(g_0) = 0$. For every g in the same coset as g_0 (i.e., $\psi(g) = \psi(g_0)$), there is a $k \in \mathcal{K}$ such that $g + k = g_0$. Since $\pi(g) + \pi(k) = \pi(g_0)$ and $\lambda(k) = \lambda(g_0) = 0$, $\lambda(g) = 0$ must also hold. Now for every g and h such that $\psi(g) = \psi(h)$, there is some $k \in \mathcal{G}$ such that $g + k = g_0$ and $h + k$ is in the same coset as $g + k$. Hence, $\lambda(g) + \lambda(k) = 0$ and $\lambda(h) + \lambda(k) = 0$. Thus, $\lambda(g) = \lambda(h)$, i.e. the λ are the same for every coset of $\mathcal{G}/\mathcal{K}$. Therefore, using that value of λ for the corresponding $h \in \mathcal{H}$ shows that the rows of $\psi(S)$, restricted to the columns where equality holds, are also linearly dependent, a contradiction.

Exercise 2*. Gomory [6, Appendix 3] gives a way of generating some faces of $\mathcal{P}(\mathcal{G}, \mathcal{N}, g_0)$ which were later studied by Devine and Glover [2]. Lately, the notion of the "core" of a facet has been studied, e.g. for the independent node problem [8]. Here, the core of a facet π is π restricted to a subset $\mathcal{S} \subseteq \mathcal{G}$ such that $\pi(g), g \notin \mathcal{S}$ can be generated one element at a time, but the same statement is not true for any proper subset $\mathscr{S}' \subset \mathscr{S}$. Try to characterize cores of facets.

Exercise 3*. Gomory [6, Thm. 21] gives a theorem with a long proof showing how to get facets from facets of $\mathscr{H} = \mathscr{G}/\mathscr{K}$ and facets of $\mathscr{K}$ in the case $g_0 \in \mathscr{K}$. Try to use a proof like that of Theorem 3 to show this result. Can the theorem be generalized to generate other facets? Is it possible to generate all of the facets of particular groups from subgroups?

REFERENCES

[1] J. ARAOZ, *Polyhedral neopolarities*, Ph.D. thesis, Dept. of Computer Sciences and Applied Analysis, Univ. of Waterloo, Waterloo, Ontario, December 1973.
[2] M. DEVINE AND F. GLOVER, *Generating the nested faces of the Gomory polyhedron*, Research Rep. R-69-1, School of Industrial Engineering, Norman, Oklahoma, December 1969.
[3] D. R. FULKERSON, *Blocking polyhedra*, Graph Theory and its Applications, B. Harris, ed., Academic Press, New York, 1970, pp. 93–112.
[4] R. E. GOMORY, *On the relation between integer and non-integer solutions to linear programs*, Proc. Nat. Acad. of Sci. U.S.A., 53 (1965), pp. 260–265.
[5] ———, *Faces of an integer polyhedron*, Ibid., 57 (1967), pp. 16–18.
[6] ———, *Some polyhedra related to combinatorial problems*, Linear Algebra and Appl., 2 (1969), pp. 451–558.
[7] E. L. JOHNSON, *Support functions, blocking pairs, and anti-blocking pairs*, Math. Programming Stud., 8 (1978), pp. 167–196.
[8] M. W. PADBERG, *On the complexity of set packing polyhedra*, Studies in Integer Programming, Ann. Discrete Math., 1 (1977), pp. 421–434.
[9] J. TIND, *Blocking and anti-blocking sets*, Math. Programming, 6 (1974), pp. 157–166.
[10] W. W. WHITE, *On a group theoretic approach to linear integer programming*, ORC 66-27, Univ. of California, Berkeley, CA, September 1966.

CHAPTER VI

Araoz's Semigroup Problem

1. Semigroups. A (finite, Abelian) *semigroup* is defined to be a finite set $\mathcal{G}$ together with addition $\hat{+}$ such that for all g, h, and k in $\mathcal{G}$:

(1) $\quad\quad\quad\quad g \hat{+} h \in \mathcal{G}\quad$ (closure);

(2) $\quad\quad\quad\quad g \hat{+} (h \hat{+} k) = (g \hat{+} h) \hat{+} k\quad$ (associative);

(3) $\quad\quad\quad\quad g \hat{+} h = h \hat{+} g\quad$ (commutative);

and

(4) $\quad\quad$ there is an element $\hat{0} \in \mathcal{G}$ such that $g \hat{+} \hat{0} = g$, for all $g \in \mathcal{G}$.

We call $\hat{0}$ the *zero* element of $\mathcal{G}$.

The first two axioms are always assumed for semigroups. Axiom (3) defines the semigroup as an Abelian semigroup. Axiom (4) is not particularly important since if $\mathcal{G}$ did not have a $\hat{0}$, then $\hat{0}$ could be adjoined to $\mathcal{G}$ with $\hat{0} \hat{+} g = g$ and $\hat{0} \hat{+} \hat{0} = \hat{0}$. Evidently, if $\mathcal{G}$ has a $\hat{0}$ then there can be only one such $\hat{0}$. We denote the nonzero elements of $\mathcal{G}$ by $\mathcal{G}_+$.

Example 1. Consider the integers $\{0, 1, \cdots, b\}$ with addition defined as

$$i \hat{+} j = \begin{cases} i + j & \text{if } i + j \leq b, \\ b & \text{if } i + j > b. \end{cases}$$

Verify that this addition defines a semigroup. Let us denote it by $\mathcal{H}_b^{\geq}$.

Example 2. Consider the set $\{0, 1, \cdots, b, \infty\}$ of integers $0, 1, \cdots, b$ together with a symbol ∞ with addition defined by

$$i \hat{+} j = \begin{cases} i + j, & i + j \leq b, \\ \infty, & i + j > b, \end{cases}$$

$$i \hat{+} \infty = \infty,$$

$$\infty \hat{+} \infty = \infty.$$

Verify the semigroup axioms for this case, which we denote by $\mathcal{H}_b^{=}$.

Define a *sub-semigroup* $\mathcal{H}$ of $\mathcal{G}$ to be a subset $\mathcal{H}$ of $\mathcal{G}$, with the same addition as $\mathcal{G}$, such that $\mathcal{H}$ is itself a semigroup. In order for such an $\mathcal{H}$ to be a semigroup, it suffices to have closure, i.e., $h_1 \in \mathcal{H}$, $h_2 \in \mathcal{H}$ implies $h_1 \hat{+} h_2 \in \mathcal{H}$, since (2) and (3) carry over from $\mathcal{G}$. For $\mathcal{S} \subseteq \mathcal{G}$, define the *semigroup generated by* $\mathcal{S}$ to be

$$\left\{ h \,\middle|\, h = \sum_{g \in \mathcal{S}} gt(g),\ t(g) \geq 0 \text{ and integer} \right\}.$$

2. The semigroup problem.

For a semigroup $\mathcal{G}$, subset $\mathcal{N} \subseteq \mathcal{G}_+$, and $g_0 \in \mathcal{G}_+$, the *semigroup* problem is

(5) $\qquad\qquad$ minimize $\sum_{g \in \mathcal{N}} c(g)t(g)$, subject to

(6) $\qquad\qquad t(g) \geq 0$ and integer, $\qquad g \in \mathcal{N}$,

(7) $\qquad\qquad \sum_{g \in \mathcal{N}} gt(g) = b$.

In writing $\sum gt(g)$, axioms (1), (2), and (3) make it unnecessary to specify the order of addition.

Example 3. Consider the knapsack problem constraints:

$$3x_1 + 7x_2 + 9x_3 \leq 21;$$

$$x_j \geq 0 \text{ and integer}, \qquad j = 1, 2, 3.$$

The associated semigroup formulation has constraint

$$1t(1) + 3t(3) + 7t(7) + 9t(9) = 21$$

in $\mathcal{K}_{21}^=$ where $x_1 = t(3)$, $x_2 = t(7)$, $x_3 = t(9)$, and the slack variable $s = t(1)$.

We have already made one assumption about $\mathcal{G}$ and $\mathcal{N}$ without loss of generality: $\hat{0} \notin \mathcal{N}$. Several more such assumptions will be made. In general, we follow Araoz [1].

If b does not belong to the sub-semigroup generated by $\mathcal{N}$, then there is no solution t to (6) and (7). Hence, assume b belongs to the sub-semigroup generated by $\mathcal{N}$.

If $\mathcal{G}$ contains any elements not in the sub-semigroup generated by $\mathcal{N}$, then those elements are not equal to any of

$$\sum_{g \in \mathcal{N}} gt(g)$$

for t as in (3), $g \in \mathcal{N}$. We can change $\mathcal{G}$ to be equal to the sub-semigroup generated by $\mathcal{N}$ without changing the problem. We, therefore, assume, henceforth, that

$$\mathcal{G} = \left\{ h = \sum_{g \in \mathcal{N}} gt(g) \mid t(g) \geq 0 \text{ and integer} \right\}.$$

If there is any $h \in \mathcal{N}$ such that $h \hat{+} g \neq g_0$ for all $g \in \mathcal{G}$, then $t(h) = 0$ in any solution t to (6) and (7). This fact is easily proven since if there were such a t, then

$$g_0 = \sum_{g \in \mathcal{N}} gt(g) = h \hat{+} \sum_{g \in \mathcal{N}} g\hat{t}(g),$$

where

$$\hat{t}(g) = \begin{cases} t(g), & g \neq h, \\ t(h) - 1, & g = h. \end{cases}$$

Now $\hat{g} \in \mathcal{G}$ where

$$\hat{g} = \sum_{g \in \mathcal{N}} g\hat{t}(g)$$

and

$$g_0 = h \hat{+} \hat{g}.$$

Thus, a contradiction is reached, and $t(h)$ must be equal to zero in any solution. Therefore, we can, without affecting the solution set in any meaningful way, delete h from $\mathcal{N}$. In other words, we assume that for each $h \in \mathcal{N}$, the set

$$\{g \in \mathcal{G} \mid g \hat{+} h = g_0\}$$

is nonempty.

Although every $h \in \mathcal{N}$ has at least one $g \in \mathcal{G}$ such that $h \hat{+} g = g_0$, there may well be $h \in \mathcal{G}$ such that $h \hat{+} g \neq g_0$ for all $g \in \mathcal{G}$. For the purposes of problem (6) and (7), all of these $h \in \mathcal{G}$ can be condensed to one element, denoted ∞, called the *infeasible element*. This infeasible element ∞ satisfies

$$g \hat{+} \infty = \infty \quad \text{for all } g \in \mathcal{G}.$$

Let us define

$$H = \{h \in \mathcal{G} \mid h \hat{+} g \neq g_0 \text{ for all } g \in \mathcal{G}\}$$

and, first, show that for all $h \in H$ and $g \in \mathcal{G}$,

$$h \hat{+} g \in H.$$

Otherwise, there is a $k \in \mathcal{G}$ such that

$$(h \hat{+} g) \hat{+} k = g_0.$$

But then $h \hat{+} (g \hat{+} k) = g_0$ where $g \hat{+} k \in \mathcal{G}$, contradicting $h \in H$. Note that $\hat{0} \notin H$ since

$$\hat{0} \hat{+} g_0 = g_0.$$

We can, thus, condense H to a single element ∞ and get a new semigroup satisfying

$$g \hat{+} \infty = \infty,$$

for all g.

3. Master semigroup problems. A *master semigroup problem* is one where $\mathcal{G}$ is any semigroup and $\mathcal{N} = \mathcal{G} - \{\hat{0}, \infty\}$. The distinguished element ∞ may or may not be in $\mathcal{G}$ depending on whether any sum $g \hat{+} h$ for g and h in $\mathcal{N}$ is equal to it. For a master problem, $b \in \mathcal{N}$ and for every $g \in \mathcal{N}$, $g \neq b$, there are one or more $h \in \mathcal{N}$ such that $g \hat{+} h = b$.

Define $\mathcal{G}_I = \mathcal{G} - \{\hat{0}, \infty\}$ and

(8) $\quad \mathcal{Q}(\mathcal{G}, b) = \text{conv} \{(t(g), g \in \mathcal{G}_I) \mid t(g) \geq 0, \text{ integer, and } \sum_{g \in \mathcal{G}} gt(g) = b\}$

and

(9) $$\mathcal{P}(\mathcal{G}, b) = \mathcal{Q}(\mathcal{G}, b) + \mathcal{R}_+^n.$$

We first ask when $\mathcal{Q}(\mathcal{G}, b) = \mathcal{P}(\mathcal{G}, b)$. Another definition is needed. For $g \in \mathcal{G}_I$ consider the sequence

$$\hat{0}, g, 2g, 3g, \cdots, kg, \cdots.$$

Since $\mathcal{G}$ is finite, there can be only a finite number of different elements here. Let

$$h_0 = kg$$

be the first occurrence of any element appearing for the second time in the sequence. Define the *order* of g to be k. Now, h_0 appears earlier in the sequence, say

$$h_0 = lg, \quad l < k.$$

The sequence of distinct semigroup elements

$$h_0 = lg, \quad h_1 = (l+1)g, \cdots, h_{k-1-l} = (k-1)g$$

is the same as

$$kg, (k+1)g, \cdots, (2k-l-1)g$$

and, in fact, repeats itself indefinitely in the sequence $\hat{0}, g, 2g, 3g, \cdots$. The sequence

$$h_0, h_1, \cdots, h_{k-1-l}$$

is called the *loop* of g, and $k - l$ is called the *loop order* of g. Define g to be a *loop element* of $\mathcal{G}$ if g belongs to its loop; i.e., $l \leq 1$.

THEOREM 1 (Araoz [1, Thm. 4.2.16]). *If b is a loop element, then $\mathcal{P}(\mathcal{G}, b) = \mathcal{Q}(\mathcal{G}, b)$.*

Proof. Suppose b is a loop element of loop order $k - 1$. Then $b = kb = 2kb = \cdots = ikb$. Hence, t given by

$$t^i(g) = \begin{cases} ik, & g = b, \\ 0, & g \neq b \end{cases}$$

is a solution for all i.

To show that $\mathcal{P}(\mathcal{G}, b) = \mathcal{Q}(\mathcal{G}, b)$, we need to exhibit a sequence t^i of solutions which are fixed for all $g \neq h$ and become large for $g = h$. We have shown such a sequence for $h = b$ but need to do so for all $h \in \mathcal{G}_I$.

For $h \neq b$, let h have order m and

$$lh = mh, \quad 0 \leq l < m.$$

Now, $h' = lh$ is not $h' = \infty$ because then

$$b = lb = l(h \,\hat{+}\, \hat{h}) = lh \,\hat{+}\, l\hat{h} = \infty \,\hat{+}\, l\hat{h} = \infty,$$

where $\hat{h}$ is such that $h \hat{+} \hat{h} = b$. Hence, $h' \neq \infty$, and for some $h'' \in \mathcal{G}$, $h' \hat{+} h'' = b$. Now,

$$b = h'' \hat{+} \hat{h}' = h'' \hat{+} lh \hat{+} h'' \hat{+} mh = h'' \hat{+} (m + i(m - l))h$$

for all $i \geq -1$. Hence, t given by

$$t^i(g) = \begin{cases} 1, & g = h'', \\ k + i(k - l), & g = h, \\ 0 & \text{otherwise} \end{cases}$$

is a sequence of solutions with $t^i(h)$ becoming large, as required.

THEOREM 2. (Maurice Queyranne gave the proof of this result, stated as a conjecture in the original lecture notes). *If $\mathcal{P}(\mathcal{G}, b) = \mathcal{Q}(\mathcal{G}, b)$, then b is a loop element.*

Proof. Suppose b is not a loop element. We then show that the loop of b must be ∞, i.e., $ib = \infty$ for some $i \geq 2$. When $ib = \infty$ for some $i \geq 2$, then $t(b) \leq i$ for any solution t because otherwise

$$\sum gt(g) = \infty.$$

Thus, it remains to show that the loop of b is ∞. Suppose not. Then, the loop of b does not contain b but does contain some $h \in \mathcal{G}$, $h \neq \hat{0}$ or ∞. For this h and some $k \geq 2$, $h = kb$. Let $\bar{h}$ be such that $h \hat{+} \bar{h} = b$. For l equal to the loop order of b.

$$h \hat{+} lb = h, \quad \text{or}$$

$$\bar{h} \hat{+} h \hat{+} lb = h \hat{+} \bar{h}, \quad \text{or}$$

$$(l + 1)b = b,$$

contradicting b not being a loop element.

This completes the proof and shows that if $g \in \mathcal{G}$ is not a loop element and if h is in the loop of g, then there does not exist any $\bar{h} \in \mathcal{G}$ such that $h \hat{+} \bar{h} = g$.

REFERENCES

[1] J. ARAOZ, *Polyhedral neopolarities,* Ph.D. thesis, Dept. of Computer Sciences and Applied Analysis, Univ. of Waterloo, Waterloo, Ontario, December 1973.

CHAPTER VII

Blockers and Polars for Master Semigroup Problems

1. The blocker of master semigroup problems. For a master semigroup problem with constraints (6) and (7) of Chapter VI, where $\mathcal{N} = \mathcal{G} - \{\hat{0}, \infty\}$, the blocker of the solutions is

(1) $\mathcal{B}(\mathcal{G}, b) = \Big\{(\pi(g), g \in \mathcal{N}) \mid \pi(g) \geq 0, g \in \mathcal{N}, \text{ and}$

$$\sum_{g \in \mathcal{N}} \pi(g) t(g) \geq 1, \text{ all } t \in \mathcal{Q}(\mathcal{G}, b)\Big\},$$

where $\mathcal{Q}(\mathcal{G}, b)$ is given by (8) of Chapter VI. Note that $\mathcal{B}(\mathcal{G}, b)$ is the same whether $\mathcal{Q}(\mathcal{G}, b)$ or $\mathcal{P}(\mathcal{G}, b)$ is used because $\pi(g) \geq 0$ has been explicitly assumed.

Example 1. Consider the set covering semigroup (Araoz [1]):

$$\mathcal{S}_m^{\geq} = \{(\beta_1, \cdots, \beta_m) \mid \beta_i = 0 \text{ or } 1\}$$

with addition

$$(\alpha_1, \cdots, \alpha_m) \mathbin{\hat{+}} (\beta_1, \cdots, \beta_m) = (\gamma_1, \cdots, \gamma_m),$$

where

$$\gamma_i = \begin{cases} 0, & \alpha_i = \beta_i = 0, \\ 1 & \text{otherwise.} \end{cases}$$

Let the right hand side $b - (1, \cdots, 1)$. Then,

$$\sum_g gt(g) = b,$$

$t(g) \geq 0$ and integer,

is precisely the set covering problem.

For $m = 2$, the problem is

$$(1, 0)t(1, 0) + (0, 1)t(0, 1) + (1, 1)t(1, 1) = (1, 1)$$

and has solutions $(1, 1, 0)$ and $(0, 0, 1)$.

Figure 1 shows the polyhedron $\mathcal{P}(\mathcal{G}, b)$ ($= \mathcal{Q}(\mathcal{G}, b)$).

Example 2. This time let the semigroup be the set partitioning semigroup:

$$\mathcal{S}_m^{=} = \{\infty\} \cup \{(\beta_1, \cdots, \beta_m) \mid \beta_i = 0 \text{ or } 1\}$$

42   CHAPTER VII

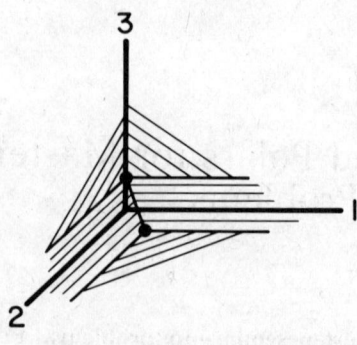

Fig. 1

with addition:

$\infty \hat{+} \infty = \infty;$

$(\alpha_1, \cdots, \alpha_m) \hat{+} \infty = \infty;$

$(\alpha_1, \cdots, \alpha_m) \hat{+} (\beta_1, \cdots, \beta_m) = \infty$ if any $\alpha_i = \beta_i = 1;$

$(\alpha_1, \cdots, \alpha_m) \hat{+} (\beta_1, \cdots, \beta_m) = (\alpha_1 + \beta_1, \cdots, \alpha_m + \beta_m)$ otherwise.

Then $\mathcal{N} = \mathcal{S}_m^= - \{\hat{0}, \infty\}$ and

$$\sum_{g \in \mathcal{N}} gt(g) = b, \quad t(g) \geq 0 \text{ and integer,}$$

is the master set partitioning problem where $b = (1, \cdots, 1)$.

For $m = 3$, the problem has group elements $\mathcal{N}$:

$$\begin{pmatrix}1\\1\\0\end{pmatrix}, \begin{pmatrix}0\\1\\0\end{pmatrix}, \begin{pmatrix}1\\1\\0\end{pmatrix}, \begin{pmatrix}0\\0\\1\end{pmatrix}, \begin{pmatrix}1\\0\\1\end{pmatrix}, \begin{pmatrix}0\\1\\1\end{pmatrix}, \begin{pmatrix}1\\1\\1\end{pmatrix}$$

and solutions (1, 1, 0, 1, 0, 0, 0), (1, 0, 0, 0, 0, 1, 0), (0, 1, 0, 0, 1, 0, 0), (0, 0, 1, 1, 0, 0, 0), (0, 0, 0, 0, 0, 0, 1). The blocker is:

```
1 0 1 0 1 0 1
0 1 1 0 0 1 1
0 0 0 1 1 1 1
```

1	1	1	1				1	≥ 1
2	1	1		1			1	≥ 1
3	1		1		1		1	≥ 1
4	1				1	1	1	≥ 1
5		1	1			1	1	≥ 1
6		1		1	1		1	≥ 1
7				1	1	1	1	≥ 1

It can be shown that the blocker can be represented by

$\pi(g) \geq 0, \quad g \in \mathcal{N},$

$\pi \cdot t \geq 1$ for all solutions t with $t(g) \leq O(g),$

BLOCKERS AND POLARS FOR MASTER SEMIGROUP PROBLEMS 43

where $O(g) \leq |\mathcal{G}|$ is the order of g. To see this result, if t is a solution with $t(h) > O(h)$ for some $h \in \mathcal{N}$, then there is a $t'(h) < t(h)$ such that $ht'(h) = ht(h)$, so $s(g)$, $g \in \mathcal{N}$, given by

$$s(g) = \begin{cases} t'(h), & g = h, \\ t(g), & g \neq h, g \in \mathcal{N} \end{cases}$$

is also a solution. Now, $\pi \cdot t \geq 1$ is implied by

$$\pi \geq 0, \qquad \pi \cdot s \geq 1.$$

Hence, $\pi \cdot t \geq 1$ was not needed to define the blocker.

Exercise 1. Generalize the notion of irreducible (Chapter IV, § 3) to sharpen the above result and so that it reduces to the previous definition in the group case. Can you prove that all vertices are irreducible in certain cases; e.g. when all $O(g) = 2$? In the group case, Gomory [2, Thm. 23] shows such a result.

2. The polar of master semigroup problems. For a master semigroup problem, the plus polar of the solutions is

(2) $$\mathcal{D}_+(\mathcal{G}, b) = \left\{ (\pi(g), g \in \mathcal{N}) \,\Big|\, \sum_{g \in \mathcal{N}} \pi(g) t(g) \geq 1, \text{ all } t \in \mathcal{Q}(\mathcal{G}, b) \right\},$$

where $\mathcal{Q}(\mathcal{G}, b)$ is given by (8). In studying $\mathcal{D}_+(\mathcal{G}, b)$, the cone

(3) $$\mathcal{D}_0(\mathcal{G}, b) = \left\{ (\pi(g), g \in \mathcal{N}) \,\Big|\, \sum_{g \in \mathcal{N}} \pi(g) t(g) \geq 0, \text{ all } t \in \mathcal{Q}(\mathcal{G}, b) \right\}$$

plays somewhat the same role as $\mathcal{R}_n^+$ since

$$\mathcal{D}_0(\mathcal{G}, b) \supseteq \mathcal{D}_+(\mathcal{G}, b) = \mathcal{D}_+(\mathcal{G}, b) + \mathcal{D}_0(\mathcal{G}, b).$$

More precisely, if the facets defining $\mathcal{D}_0(\mathcal{G}, b)$,

$$\mathcal{D}_0(\mathcal{G}, b) = \left\{ (\pi(g), g \in \mathcal{N}) \,\Big|\, \sum_{g \in \mathcal{N}} \pi(g) r^k(g) \geq 0, k = 1, \cdots, K \right\}$$

for a finite set $(r^k(g), g \in \mathcal{N})$, are known, then $\mathcal{D}_+(\mathcal{G}, b)$ can be defined by the inequality system

$$\sum_{g \in \mathcal{N}} \pi(g) r^k(g) \geq 0, \qquad k = 1, \cdots, K,$$

$$\sum_{g \in \mathcal{N}} \pi(g) t^l(g) \geq 1, \qquad l = 1, \cdots, L,$$

for the finite set $(t^l(g), g \in \mathcal{N})$ which are minimal with respect to $\mathcal{D}_0(\mathcal{G}, b)$; i.e., those $t(g)$ which are not equal to $s + r$ for some $s \in \mathcal{D}_+(\mathcal{G}, b)$ and $r \in \mathcal{D}_0(\mathcal{G}, b)$, $r \neq 0$.

It should be clear that $\mathcal{D}_0(\mathcal{G}, b) \subseteq \mathcal{R}_+^n$ since every $t \geq 0$ in a solution. We next give conditions for $\pi(g) \geq 0$ to be true for $\pi \in \mathcal{D}_0(\mathcal{G}, b)$, i.e., $\mathcal{D}_0(\mathcal{G}, b) = \mathcal{R}_+^n$.

THEOREM 1. $\mathcal{D}_0(\mathcal{G}, b) = \mathcal{R}_+^n$ *if and only if we do not have that both of the conditions below hold for h:*

(i) *the loop of h is the single element ∞; and*
(ii) $ht(h) \neq b$ *for* $t(h) = 1, 2, \cdots$.

Proof. Suppose both (i) and (ii) hold for h. Then the loop of h is ∞, so $ht(h) = \infty$ for all $t(h) \geq k$, some $k \geq 2$. By $g \mathbin{\hat{+}} \infty = \infty$,

$$\sum_{g \in \mathcal{N}} gt(g) = \infty$$

for any t having $t(h) \geq k$. Hence, $t(h) < k$ for all solutions t. Furthermore, every solution t has

$$\sum_{\substack{g \in \mathcal{N} \\ g \neq h}} t(g) \geq 1$$

because $ht(h) \neq b$ for any $t(h)$. Hence, π defined by

$$\pi(g) = \begin{cases} 1, & g \neq h, \\ -\dfrac{1}{k}, & g = h \end{cases}$$

satisfies $\pi \cdot t \geq 0$ for all solutions t. Therefore, $\pi(h) \geq 0$ is not satisifed for all $\pi \in \mathcal{D}_0(\mathcal{G}, b)$.

Conversely, suppose that $ht(h) = b$ for some nonnegative integer $t(h)$ or the loop of h is not ∞. First, if $ht(h) = b$, then setting $t(g) = 0$, $g \neq h$, gives the inequality

$$\pi(h)t(h) \geq 0$$

or

$$\pi(h) \geq 0.$$

So suppose no $ht(h) = b$, but the loop of h is not ∞. Since $g \mathbin{\hat{+}} \infty = \infty$, for all $g \in \mathcal{G}$, it must be true that $ht(h) \neq \infty$ for any $t(h)$. Now, let $h' \in \mathcal{N}$ be in the loop of h. Then, $h' = ih$ for some integer $i \geq 1$, and $h' = (i + m(l + 1))h$, where l is the loop order of h. By assumption, there is some $h'' \in \mathcal{N}$ such that $h' \mathbin{\hat{+}} h'' = b$. Hence,

$$(i + m(l + 1))h \mathbin{\hat{+}} h'' = b,$$

so

$$(i + m(l + 1))\pi(h) + \pi(h'') \geq 0.$$

Any π having $\pi(h) < 0$ would violate the above inequality for some m large enough. Hence, $\pi(h) \geq 0$ must hold for all $\pi \in \mathcal{D}_0(\mathcal{G}, b)$.

Exercise 2. For the semigroup with elements

$$\binom{1}{0}, \binom{2}{0}, \binom{3}{0}, \binom{1}{1}, \binom{2}{1}, \binom{3}{1}, \infty$$

with addition defined by

$$\begin{pmatrix} i_1 \\ i_2 \end{pmatrix} \dotplus \begin{pmatrix} j_1 \\ j_2 \end{pmatrix} = \begin{cases} \infty & \text{if } i_2 + j_2 \geq 2, \\ \begin{pmatrix} 2 + (i_1 + j_1 - 2 \pmod 2) \\ i_2 + j_2 \end{pmatrix} & \text{otherwise,} \end{cases}$$

find $\mathcal{D}_0(\mathcal{G}, b)$, where $b = \begin{pmatrix} 3 \\ 1 \end{pmatrix}$. Give the order of each element and the loop order of each.

From Theorem 1, we can define $\mathcal{D}_0(\mathcal{G}, b)$ by the finite system

$\pi(g) \geq 0$ where implied by Theorem 2,

$\pi \cdot t \geq 0$ for all solutions t satisfying $t(g) \leq O(g)$.

The inequalities $\pi \cdot t \geq 0$ need be written only for t having $t(h) \geq 1$ for some h for which $\pi(h) \geq 0$ cannot be written.

From the above, the polyhedron $\mathcal{D}_+(\mathcal{G}, b)$ is equal to the solutions π to

$\pi(g) \geq 0$ where implied by Theorem 2,

$\pi(t) \geq 1$ for all solutions t satisfying $t(g) \leq O(g)$.

Example 3. Consider the problem

$$\begin{pmatrix} 1 \\ 0 \end{pmatrix} t \begin{pmatrix} 1 \\ 0 \end{pmatrix} + \begin{pmatrix} 0 \\ 1 \end{pmatrix} t \begin{pmatrix} 0 \\ 1 \end{pmatrix} + \begin{pmatrix} 1 \\ 1 \end{pmatrix} t \begin{pmatrix} 1 \\ 1 \end{pmatrix} = \begin{pmatrix} 1 \\ 1 \end{pmatrix},$$

where addition is defined for $g_1 = \begin{pmatrix} 1 \\ 0 \end{pmatrix}$, $g_2 = \begin{pmatrix} 0 \\ 1 \end{pmatrix}$, $g_3 = \begin{pmatrix} 1 \\ 1 \end{pmatrix}$ by

	g_1	g_2	g_3	∞
g_1	g_1	g_3	g_3	∞
g_2	g_3	∞	∞	∞
g_3	g_3	∞	∞	∞
∞	∞	∞	∞	∞

Here, g_1 has order 2, g_2 and g_3 have order 3. $\mathcal{D}_0(\mathcal{G}, b)$ is the cone:

$\pi(g_1) \geq 0$;

$\pi(g_1) + \pi(g_2) \geq 0$;

$\pi(g_3) \geq 0$;

and has extreme rays $(1, -1, 0)$, $(0, 1, 0)$, $(0, 0, 1)$ as shown in Fig. 2(a). The polyhedron $\mathcal{D}_+(\mathcal{G}, b)$ is:

$\pi(g_1) \geq 0$;

$\pi(g_3) \geq 1$;

and

$\pi(g_1) + \pi(g_2) \geq 1$,

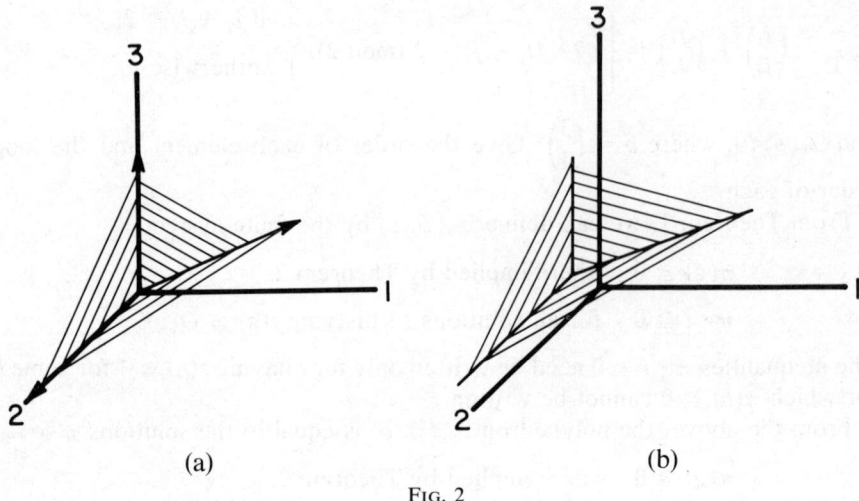

(a)　　　　　　　　　　　　　(b)

Fig. 2

and has one vertex, (0, 1, 1), and three rays, (1, −1, 0), (0, 1, 0), (1, 0, 0) as shown in Fig. 2(b). The polyhedron $\mathscr{Q}(\mathscr{G}, b)$ is shown in Fig. 3. The defining inequalities are

$$t_2 + t_3 = 1,$$
$$t_1 - t_2 \geqq 0,$$
$$t_j \geqq 0, \quad j = 1, 2, 3.$$

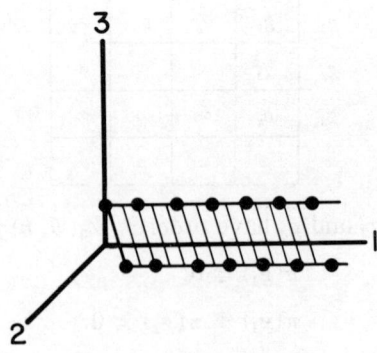

Fig. 3

REFERENCES

[1] J. Araoz, *Polyhedral neopolarities*, Ph.D. thesis, Dept. of Applied Analysis and Computer Science, Univ. of Waterloo, Waterloo, Ontario, November 1973.
[2] R. E. Gomory, *Some polyhedra related to combinatorial problems*, Linear Algebra and Appl., 2 (1969), pp. 451–558.

CHAPTER VIII

Subadditive and Minimal Valid Inequalities

1. The subadditive cone $\mathscr{S}(\mathscr{G})$. In this chapter, we concentrate on the master semigroup problem:

(1) $$t(g) \geq 0 \text{ and integer}, \quad g \in \mathcal{N},$$

(2) $$\sum_{g \in \mathscr{G}} gt(g) = b,$$

where $\mathcal{N} = \mathscr{G} - \{\hat{0}, \infty\}$, $b \in \mathcal{N}$, and $\mathscr{G}$ includes ∞ if and only if $g \hat{+} h$ equals ∞ for some pair $g, h \in \mathcal{N}$.

Define the *subadditive cone* $\mathscr{S}(\mathscr{G})$ by

(3) $\mathscr{S}(\mathscr{G}) = \{(\pi(g), g \in \mathcal{N}) \mid$ for all $g, h \in \mathcal{N}$,
$$\pi(y \hat{+} h) \leq \pi(g) + \pi(h), \text{ if } g \hat{+} h \in \mathcal{N},$$
$$0 \leq \pi(g) + \pi(h), \text{ if } g \hat{+} h = \hat{0}\}.$$

We emphasize that $\pi(g) + \pi(h)$ is not restricted if $g \hat{+} h = \infty$. Thus, the subadditive cone is defined by the addition table for $\mathscr{G}$. It is a polyhedral cone because it is defined by a finite number of homogeneous inequalities.

THEOREM 1. *The lineality of $\mathscr{S}(\mathscr{G})$ is the set of $(\pi(g), g \in \mathcal{N})$ satisfying the inequality conditions in (3) with equality.*

This result is a standard result for polyhedral cones because if $\pi \in \mathscr{S}(\mathscr{G})$ and $-\pi \in \mathscr{S}(\mathscr{G})$ then every defining inequality for $\mathscr{S}(\mathscr{G})$ must hold for π and $-\pi$ so every defining inequality must hold with equality.

Example 1. We go back to the example $\mathscr{K}_2^=$ with addition: $1 \hat{+} 1 = 2$, $1 \hat{+} 2 = \infty$, $2 \hat{+} 2 = \infty$. The cone $\mathscr{S}(\mathscr{G})$ is shown in Fig. 1. The lineality is the line $\pi_2 = 2\pi_1$.

2. Subadditive valid inequalities. Define three polyhedral sets from $\mathscr{S}(\mathscr{G})$ by

(4) $$\mathscr{S}_+(\mathscr{G}, b) = \{(\pi(g), g \in \mathcal{N}) \in \mathscr{S}(\mathscr{G}) \mid \pi(b) \geq 1\},$$

(5) $$\mathscr{S}_0(\mathscr{G}, b) = \{(\pi(g), g \in \mathcal{N}) \in \mathscr{S}(\mathscr{G}) \mid \pi(b) \geq 0\},$$

(6) $$\mathscr{S}_-(\mathscr{G}, b) = \{(\pi(g), g \in \mathcal{N}) \in \mathscr{S}(\mathscr{G}) \mid \pi(b) \geq -1\}.$$

The polyhedron $\mathscr{S}_0(\mathscr{G}, b)$ is still a cone, but the other two are not cones because the conditions $\pi(b) \geq 1$ and $\pi(b) \geq -1$ are not homogeneous.

In Chapter VII, we defined $\mathscr{D}_+(\mathscr{G}, b)$ and $\mathscr{D}_0(\mathscr{G}, b)$. Now, $\mathscr{D}_-(\mathscr{G}, b)$ will be introduced, and all three are defined below.

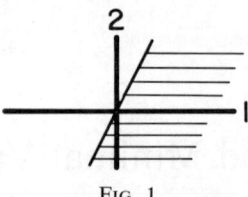

Fig. 1

(7) $\mathcal{D}_+(\mathcal{G}, b) = \left\{(\pi(g), g \in \mathcal{N}) \mid \sum_{g \in \mathcal{N}} \pi(g)t(g) \geq 1, \text{ all } t \in \mathcal{Q}(\mathcal{G}, b)\right\}$,

(8) $\mathcal{D}_0(\mathcal{G}, b) = \left\{(\pi(g), g \in \mathcal{N}) \mid \sum_{g \in \mathcal{N}} \pi(g)t(g) \geq 0, \text{ all } t \in \mathcal{Q}(\mathcal{G}, b)\right\}$,

(9) $\mathcal{D}_-(\mathcal{G}, b) = \left\{(\pi(g), g \in \mathcal{N}) \mid \sum_{g \in \mathcal{N}} \pi(g)t(g) \geq -1, \text{ all } t \in \mathcal{Q}(\mathcal{G}, b)\right\}$.

Clearly, $\mathcal{D}_+(\mathcal{G}, b) \subseteq \mathcal{D}_0(\mathcal{G}, b) \subseteq \mathcal{D}_-(\mathcal{G}, b)$.

Example 2. We use, again, $\mathcal{K}_2^=$. Figure 2 shows the three sets. Here, $\mathcal{Q}(\mathcal{G}, b)$ is the convex hull of $(2, 0)$, $(0, 1)$. The two sets $\mathcal{D}_+$ and $\mathcal{D}_-$ have one extreme point each: $(\tfrac{1}{2}, 1)$, $(-\tfrac{1}{2}, -1)$ giving two inequalities

$$\tfrac{1}{2}t_1 + t_2 \geq 1 \quad \text{or} \quad t_1 + 2t_2 \geq 2$$

and

$$-\tfrac{1}{2}t_1 - t_2 \geq -1 \quad \text{or} \quad t_1 + 2t_2 \geq 2.$$

The cone $\mathcal{D}_0$, which is the recession cone of both $\mathcal{D}_+$ and $\mathcal{D}_-$, has only the extreme rays corresponding to the axis directions and giving $t_1 \geq 0$, $t_2 \geq 0$.

The three sets $\mathcal{D}_+(\mathcal{G}, b)$, $\mathcal{D}_0(\mathcal{G}, b)$, and $\mathcal{D}_-(\mathcal{G}, b)$ give valid inequalities such that $\mathcal{Q}(\mathcal{G}, b)$ is the intersection of them all. The valid inequalities can be obtained from $\mathcal{D}_+(\mathcal{G}, b)$ and $\mathcal{D}_-(\mathcal{G}, b)$ by taking their extreme points and from $\mathcal{D}_0(\mathcal{G}, b)$ by taking its extreme rays. The valid equations are of two types: π in the lineality of $\mathcal{D}_0(\mathcal{G}, b)$, or π such that $\pi \in \mathcal{D}_+(\mathcal{G}, b)$ and $-\pi \in \mathcal{D}_-(\mathcal{G}, b)$.

The three sets satisfy

$$\mathcal{D}_+(\mathcal{G}, b) + \mathcal{D}_-(\mathcal{G}, b) \subseteq \mathcal{D}_0(\mathcal{G}, b).$$

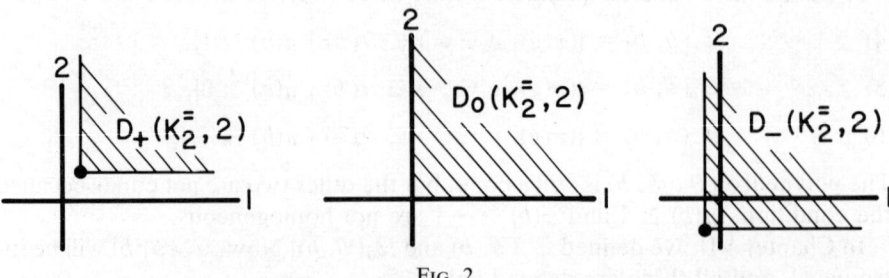

Fig. 2

In fact, that condition, along with the fact that $\mathscr{D}_0(\mathscr{G}, b)$ is the recession cone of both $\mathscr{D}_+(\mathscr{G}, b)$ and $\mathscr{D}_-(\mathscr{G}, b)$ satisfying $\mathscr{D}_-(\mathscr{G}, b) \supseteq \mathscr{D}_0(\mathscr{G}, b) \supseteq \mathscr{D}_+(\mathscr{G}, b)$, characterizes such sets (see [5, Thm. 2]).

We now relate the $\mathscr{D}$'s and $\mathscr{S}$'s.

THEOREM 2. *The following three inclusions hold:*

$$\mathscr{S}_+(\mathscr{G}, b) \subseteq \mathscr{D}_+(\mathscr{G}, b);$$

$$\mathscr{S}_0(\mathscr{G}, b) \subseteq \mathscr{D}_0(\mathscr{G}, b), \text{ and}$$

$$\mathscr{S}_-(\mathscr{G}, b) \subseteq \mathscr{D}_-(\mathscr{G}, b).$$

The proof of Theorem 2 is trivial from Lemma 3 below. This development is similar to that in [3, Thm. 1.5].

LEMMA 3. *If $\pi \in \mathscr{S}(\mathscr{G})$ and if s satisfies*

$$s(g) \geq 0 \text{ and integer}, \qquad g \in \mathscr{N},$$

$$\sum_{g \in \mathscr{N}} gs(g) \in \mathscr{N} \cup \{\hat{0}\},$$

then

(10) $$\pi\left(\sum_{g \in \mathscr{N}} gs(g)\right) \leq \sum_{g \in \mathscr{N}} \pi(g)s(g),$$

where we take $\pi(\hat{0}) = 0$.

Proof. The proof is by induction on the sum

$$S(s) = \sum_{g \in \mathscr{N}} s(g).$$

For $S(s) = 1$, some $h \in \mathscr{N}$ has $s(h) = 1$ and every other $s(g) = 0$. The statement is trivial then: $\pi(h) \leq \pi(h)$.

Suppose, as induction hypothesis, that the lemma holds for all s having $S(s) \leq k$ for some $k \leq 1$. Now let $S(s) = k + 1$. Pick any h such that $s(h) \geq 1$. Let s' be given by

$$s'(g) = \begin{cases} s(g), & g \neq h, \\ s(h) - 1, & g = h. \end{cases}$$

Then,

$$\pi\left(\sum_{g \in \mathscr{N}} gs(g)\right) = \pi\left(h + \sum_{g \in \mathscr{N}} gs'(g)\right)$$

$$\leq \pi(h) + \left(\sum_{g \in \mathscr{N}} gs'(g)\right), \qquad \text{by (4)}$$

$$\leq \pi(h) + \sum_{g \in \mathscr{N}} \pi(g)s'(g), \qquad \text{by the induction hypothesis,}$$

$$= \sum_{g \in \mathscr{N}} \pi(g)s(g), \qquad \text{by definition of } s'.$$

The only consideration here is whether

$$h' = \sum_{g \in \mathcal{N}} g(s'(g))$$

is in $\mathcal{N}$ or not. Clearly $h' \neq \infty$ since $h' + h \neq \infty$. If $h' = \hat{0}$, then $\pi(h + h') \leq \pi(h) + \pi(h')$ is true by $h + h' = h$ and $\pi(h') = 0$. The induction hypothesis was stated for $N \cup \{\hat{0}\}$ so can still be used, and the proof goes through. If $h' \in \mathcal{N}$, then (4) applies and so does the induction hypothesis. Hence, the proof is completed.

Define a *nonnegative valid inequality* to be any $\pi \in \mathcal{B}(\mathcal{G}, b)$, where $\mathcal{B}(\mathcal{G}, b)$ is given by (9) of Chapter IV to be valid inequalities also satisfying $\pi(g) \geq 0$, all $g \in \mathcal{N}$. Define the set $\overline{\mathcal{G}}(\mathcal{G}, b)$ of *nonnegative subadditive valid inequalities* to be those $\pi \in \mathcal{G}_+(\mathcal{G}, b)$ which also satisfy $\pi(g) \geq 0$, all $g \in \mathcal{N}$. Theorem 2 establishes that a nonnegative subadditive valid inequality is a valid inequality.

From Theorem 1 of Chapter VII, for a particular $h \in \mathcal{N}$, $\pi(h) \geq 0$ holds for all valid inequalities if and only if it is not true that both the loop of h is ∞ and $ih \neq b$ for all positive integers i. That theorem was for $\mathcal{D}_0(\mathcal{G}, b)$, i.e., $\pi \cdot t \geq 0$ rather than $\pi \cdot t \geq 1$. However, $\mathcal{D}_0(\mathcal{G}, b) \supseteq \mathcal{D}_+(\mathcal{G}, b) = \mathcal{D}_+(\mathcal{G}, b) + \mathcal{D}_0(\mathcal{G}, b)$, so the same result holds for $\mathcal{D}_+(\mathcal{G}, b)$. Since subadditive valid inequalities are valid inequalities, $\pi(h) \geq 0$ holds for all subadditive valid inequalities whenever the condition of Theorem 1 of Chapter VII holds.

Lifting the restrictions of nonnegativity on π for semigroup problems was done by Araoz [1].

3. Minimal valid inequalities. A valid inequality $(\pi(g), g \in \mathcal{N}) \in \mathcal{D}_+(\mathcal{G}, b)$ is called *minimal* if every ρ satisfying $\rho(g) \leq \pi(g)$, for all $g \in \mathcal{N}$, with $\rho(h) < \pi(h)$ for at least one $h \in \mathcal{N}$, is not a valid inequality in $\mathcal{D}_+(\mathcal{G}, b)$.

Related results to the next theorem appear in [1], [3], and [4]. Our proof is very similar to that of Theorem 1.2 of [3].

THEOREM 4. *Every minimal valid inequality is subadditive.*

Proof. Suppose that π is a minimal valid inequality and is not subadditive. Then, for some $h' \in \mathcal{N}$ and $h^2 \in \mathcal{N}$ with $h^1 + h^2 \neq \infty$,

$$\pi(h^1 + h^2) > \pi(h^1) + \pi(h^2).$$

There are two cases: $h^1 + h^2 = \hat{0}$ and $h^1 + h^2 \in \mathcal{N}$.

Case (i): $h^1 + h^2 = \hat{0}$. Then, for any positive integer i,

$$ih^1 \hat{+} ih^2 = i(h^1 + h^2) = \hat{0}.$$

For any solution t, t' given by

$$t'(g) = \begin{cases} t(g) + i, & g = h^1 \text{ or } g = h^2, \\ t(g), & \text{otherwise}, \end{cases}$$

is also a solution since

$$\sum_{g \in \mathcal{N}} gt'(g) = \sum_{g \in \mathcal{N}} gt(g) \hat{+} ih^1 \hat{+} ih^2 = \sum_{g \in \mathcal{N}} gt(g) = b.$$

However,
$$\sum_{g \in \mathcal{N}} \pi(g)t'(g) = \sum_{g \in \mathcal{N}} \pi(g)t(g) + i(\pi(h^1) + \pi(h^2))$$
can be made arbitrarily small by taking i large, since $\pi(h^1) + \pi(h^2) < \pi(h^1 + h^2) = \pi(\hat{0}) = 0$. Hence, π is not valid, and a contradiction is reached.

Case (ii): $h^1 + h^2 \in \mathcal{N}$. Define π' by
$$\pi'(g) = \begin{cases} \pi(h^1) + \pi(h^2), & g = h^1 + h^2, \\ \pi(g), & \text{otherwise.} \end{cases}$$

If π' is a valid inequality, then a contradiction to minimality of π is reached. Suppose π' is not valid. Then, for some solution t,
$$\pi' \cdot t < 1.$$

Define t' from t by
$$t'(g) = \begin{cases} 0, & g = h^1 + h^2, \\ t(g) + t(h^1 + h^2), & g = h^1 \text{ or } h = h^2, \\ t(g), & \text{otherwise.} \end{cases}$$

Then, t' is a solution since
$$\sum_{g \in \mathcal{N}} gt'(g) = \sum_{\substack{g \in \mathcal{N} \\ g \neq h^1, h^2, h^1+h^2}} gt(g) \hat{+} h^1 t(h^1) \hat{+} h^1 t(h^1 + h^2) \hat{+} h^2 t(h^2) \hat{+} h^2 t(h^1 + h^2),$$

$$= \sum_{g \in \mathcal{N}} gt(g) = b.$$

Now,
$$\pi \cdot t' = \sum_{g \in \mathcal{N}} \pi(g)t'(g)$$

$$= \sum_{\substack{g \in \mathcal{N} \\ g \neq h^1, h^2, h^1+h^2}} \pi(g)t'(g) + \pi(h^1)t'(h^1) + \pi(h^2)t'(h^2)$$

$$\qquad + \pi(h^1 + h^2)t'(h^1 + h^2),$$

$$= \sum_{\substack{g \in \mathcal{N} \\ g \neq h^1, h^2, h^1+h^2}} \pi'(g)t(g) + \pi(h^1)(t(h^1) + t(h^1 + h^2))$$

$$\qquad + \pi(h^2)(t(h^2) + t(h^1 + h^2)),$$

$$= \sum_{\substack{g \in \mathcal{N} \\ g \neq h^1+h^2}} \pi'(g)t(g) + (\pi(h^1) + \pi(h^2))t(h^1 + h^2),$$

$$= \sum_{g \in \mathcal{N}} \pi'(g)t(g) = \pi' \cdot t < 1.$$

Thus, a contradiction to π being valid is reached, and the theorem is proven. We remark that the theorem also holds for $\mathcal{D}_0(\mathcal{G}, b)$ or $\mathcal{D}_-(\mathcal{G}, b)$ replacing $\mathcal{D}_+(\mathcal{G}, b)$.

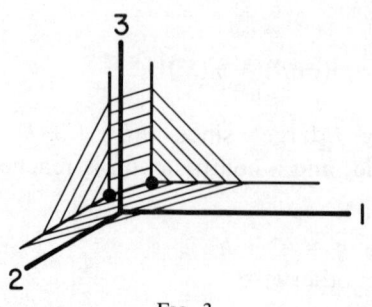

Fig. 3

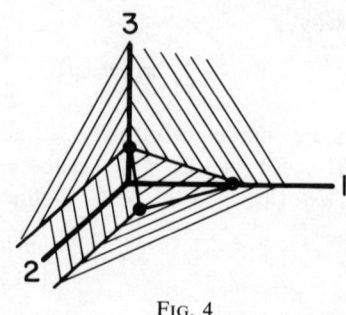

Fig. 4

Example 3. Consider $\mathcal{K}_3^=$:

$$1t_1 + 2t_2 + 3t_3 = 3,$$

$$t_j \geqq 0 \text{ and integer,}$$

where $i + j = \infty$ if $i + j > 3$. The only solutions are $(3, 0, 0), (1, 1, 0), (0, 0, 1)$. Then, $\mathcal{B}(\mathcal{K}_3^=, 3)$ is as shown in Fig. 3 and has vertices $(1, 0, 1), (\frac{1}{3}, \frac{2}{3}, 1)$ and rays $(1, 0, 0), (0, 1, 0), (0, 0, 1)$. The system of inequalities giving $\mathcal{P}(\mathcal{K}_3^=, 3)$ is, thus,

$$t_j \geqq 0,$$

$$t_1 + 2t_2 + 3t_3 \geqq 3,$$

$$t_1 + t_3 \geqq 1,$$

and is shown in Fig. 4. Now $\mathcal{D}_+(\mathcal{K}_3^=, 3)$ is shown in Fig. 5, where $\pi(2) \geqq 0$ is not imposed. It has one vertex, $(\frac{1}{3}, \frac{2}{3}, 1)$, and three rays, $(0, 1, 0), (0, 0, 1), (1, -1, 0)$.

The system of inequalities for $\mathcal{Q}(\mathcal{K}_3^=, 3)$ is

$$t_2 \geqq 0, \quad t_3 \geqq 0,$$

$$t_1 - t_2 \geqq 0,$$

$$t_1 + 2t_2 + 3t_3 = 3,$$

and is shown in Fig. 6.

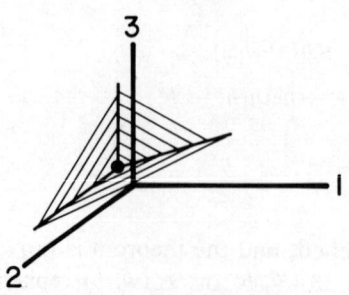

Fig. 5

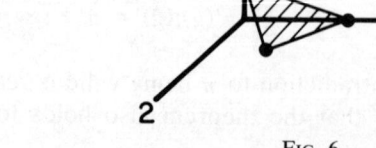

Fig. 6

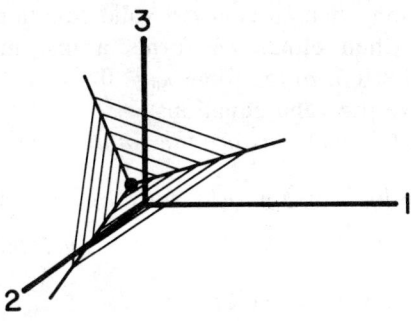

Fig. 7

The polyhedron $\mathscr{S}_+(\mathscr{G}, b)$ of subadditive valid inequalities is defined by the inequalities:

$$\pi(2) \leq 2\pi(1),$$
$$\pi(3) \leq \pi(1) + \pi(2),$$
$$\pi(3) \geq 1,$$

as shown in Fig. 7. It has one vertex, $(\tfrac{1}{3}, \tfrac{2}{3}, 1)$, and three rays, $(1, 2, 0)$, $(1, 2, 3)$, and $(1, -1, 0)$.

The minimal valid inequalities are the half line $(\tfrac{1}{3}, \tfrac{2}{3}, 1) + \lambda(1, -1, 0)$, $\lambda \geq 0$. This half line is also present in Fig. 7, illustrating Theorem 4. The minimal valid inequalities in $\mathscr{B}(\mathscr{K}_3^=, 3)$ (see Fig. 3) are the line segment $\alpha(\tfrac{1}{3}, \tfrac{2}{3}, 1) + (1 - \alpha)(1, 0, 1)$, $0 \leq \alpha \leq 1$, which is also in Fig. 7.

Theorem 2 is illustrated by the fact that the polyhedron in Fig. 7 is a subset of that in Fig. 5. If we intersect that of Fig. 7 with $\mathscr{R}_3^+$, i.e., $\pi(2) \geq 0$, then the resulting polyhedron is a subset of that of Fig. 3.

Denote the nonnegative subadditive valid inequalities by $\overline{\mathscr{S}}_+(\mathscr{G}, b)$. Theorem 2 then says $\mathscr{S}_+(\mathscr{G}, b) \subseteq \mathscr{D}_+(\mathscr{G}, b)$, and $\overline{\mathscr{S}}_+(\mathscr{G}, b) \subseteq \mathscr{B}(\mathscr{G}, b)$ also follows. Theorem 4 implies that the reverse inclusions hold if we restrict $\mathscr{D}_+(\mathscr{G}, b)$ and $\mathscr{B}(\mathscr{G}, b)$, respectively, to their minimal elements.

Exercise 1. For the set partitioning semigroup $\mathscr{S}_2^=$ with elements $\left\{ \hat{0}, \binom{1}{0}, \binom{0}{1}, \binom{1}{1}, \infty \right\}$, and $b = \binom{1}{1}$, draw $\mathscr{P}(\mathscr{S}_2^=, b)$, $\mathscr{Q}(\mathscr{S}_2^=, b)$, $\mathscr{B}(\mathscr{S}_2^=, b)$, $\mathscr{B}_+(\mathscr{S}_2^=, b)$, $\mathscr{S}_+(\mathscr{S}_2^=, b)$, and $\overline{\mathscr{S}}_+(\mathscr{S}_2^=, b)$.

4. Valid equations. The polyhedron $\mathscr{P}(\mathscr{G}, b)$ is full dimensional and so has no valid equations. Correspondingly, $\mathscr{B}(\mathscr{G}, b)$ is *pointed,* i.e., does not contain any complete lines. However, $\mathscr{Q}(\mathscr{G}, b)$ may not be full dimensional, so $\mathscr{D}(\mathscr{G}, b)$ may contain lines. Define α to be a *valid equation* for $\mathscr{Q}(\mathscr{G}, b)$ if

$$\sum_{g \in \mathscr{N}} \alpha(g) t(g) = \alpha(b)$$

holds for all $t \in \mathscr{Q}(\mathscr{G}, b)$. Every valid equation has to have a right hand side equal to $\alpha(b)$, because $t(b) = 1$, $t(g) = 0$, $g \neq b$, is obviously a solution.

If α is a valid equation, then $\lambda\alpha$ is also a valid equation for all real λ. If α is not identically zero, then either $\lambda\alpha$ forms a line in $\mathscr{D}_0(\mathscr{G}, b)$ or $\lambda_0\alpha \in \mathscr{D}_+(\mathscr{G}, b)$ and $-\lambda_0\alpha \in \mathscr{D}_-(\mathscr{G}, b)$ for some $\lambda_0 \neq 0$.

We now characterize the valid equations.

THEOREM 5. *The valid equations are the solution set of*

$$\{(\alpha(g), g \in \mathscr{N}) \mid \alpha(g + h) = \alpha(g) + \alpha(h), \text{ for all } g, h \in \mathscr{N}$$
$$\text{such that } g \hat{+} h \in \mathscr{N} \cup \{\hat{0}\}\}.$$

Proof. Suppose α is a valid equation. Then,

$$\sum_{g \in \mathscr{N}} \alpha(g) t(g) = \alpha(b)$$

for all $t \in \mathscr{Q}(\mathscr{G}, b)$, in particular, for all solutions t. Suppose α does not satisfy

$$\alpha(h^1 \hat{+} h^2) = \alpha(h^1) + \alpha(h^2)$$

for some $h^1, h^2 \in \mathscr{N}$ and $h^1 \hat{+} h^2 \in \mathscr{N} \cup \{\hat{0}\}$. If $h^1 \hat{+} h^2 = \hat{0}$, then

$$0 \neq \alpha(h^1) + \alpha(h^2).$$

For any solution t and for t' given by

$$t'(g) = \begin{cases} t(g) + 1, & g = h^1, h^2, \\ t(g), & \text{otherwise,} \end{cases}$$

t' is also a solution since

$$\sum_{g \in \mathscr{N}} gt'(g) = \sum_{g \in \mathscr{N}} gt(g) \hat{+} h^1 \hat{+} h^2 = \sum_{g \in \mathscr{N}} gt(g).$$

But,

$$\sum_{g \in \mathscr{N}} \alpha(g) t'(g) = \sum_{g \in \mathscr{N}} \alpha(g) t(g) + \alpha(h^1) + \alpha(h^2) \neq \sum_{g \in \mathscr{N}} \alpha(g) t(g),$$

contradicting α being a valid equation.

If $h^1 \hat{+} h^2 \in \mathscr{N}$, then

$$\alpha(h^1 \hat{+} h^2) \neq \alpha(h^1) + \alpha(h^2).$$

Let h^3 be such that $(h^1 \hat{+} h^2) \hat{+} h^3 = b$. Define t and t' by

$$t(g) = \begin{cases} 1, & g = h^1 \hat{+} h^2, h^3, \\ 0 & \text{otherwise,} \end{cases}$$

$$t'(g) = \begin{cases} 1, & g = h^1, h^2, h^3, \\ 0 & \text{otherwise.} \end{cases}$$

Both t and t' are solutions but

$$\sum_{g\in \mathcal{N}} \alpha(g)t(g) = \alpha(h^1 \hat{+} h^2) + \alpha(h^3)$$

$$\neq \alpha(h^1) + \alpha(h^2) + \alpha(h^3) = \sum_{g\in \mathcal{N}} \alpha(g)t'(g),$$

again contradicting α being a valid equation.

To show the converse, let α satisfy

$$\alpha(g \hat{+} h) = \alpha(g) + \alpha(h)$$

for all $g, h \in \mathcal{N}$ and $g \hat{+} h \in \mathcal{N} \cup \{\hat{0}\}$. The proof that α is a valid equation consists in showing that

$$\alpha\left(\sum_{g\in \mathcal{N}} gt(g)\right) = \sum_{g\in \mathcal{N}} \alpha(g)t(g), \quad t(g) \geq 0 \text{ and integer.}$$

This additivity of α follows much as in Lemma 2 and will not be repeated here.

Theorem 5 says that the lineality of $\mathscr{S}(\mathcal{G})$ is equal to the set of valid equations.

5. Homogeneous valid inequalities. For $\mathcal{Q}(\mathcal{G}, b)$ the inequalities $t(g) \geq 0$ are always valid, of course, but there may be other facets $\pi \cdot t \geq 0$ which are not nonnegative linear combinations of $t(g) \geq 0$; i.e., $\pi(g) < 0$ may happen for some g. In terms of $\mathcal{D}_0(\mathcal{G}, b)$, the unit vectors are always rays which can be added to any point in $\mathcal{D}_0(\mathcal{G}, b)$ without leaving $\mathcal{D}_0(\mathcal{G}, b)$. However, these rays may not be extreme and $\mathcal{R}_+^n$ may be a proper subset of $\mathcal{D}_0(\mathcal{G}, b)$.

THEOREM 6. *The extreme rays of $\mathcal{D}_0(\mathcal{G}, b)$, other than the unit vectors, are also extreme rays of $\mathscr{S}_0(\mathcal{G}, b)$ and are precisely those extreme rays of $\mathscr{S}_0(\mathcal{G}, b)$ which are minimal.*

Proof. If π is an extreme ray of $\mathcal{D}_0(\mathcal{G}, b)$ and is not a unit vector, then neither $\pi \geq 0$ nor $\pi \leq 0$ is true because if $\pi \geq 0$ then π is not extreme and if $\pi \leq 0$ then π is not valid unless $\pi = 0$.

If $\rho < \pi$ is also in the cone $\mathcal{D}_0(\mathcal{G}, b)$, then $\pi + (\pi - \rho) > \pi$ must also be in it and

$$\pi = \tfrac{1}{2}\rho + \tfrac{1}{2}(\pi + (\pi - \rho)),$$

contradicting extremality of π unless either ρ lies on the same ray as π or $\pi - \rho$ belongs to the lineality. But ρ cannot be on the same ray as π because $\rho < \pi$ and $\pi \not\geq 0$. Since $\pi - \rho > 0$, it cannot be in the lineality. Hence, π must be minimal.

If π were not subadditive, then

$$\pi(g \hat{+} h) > \pi(g) + \pi(h)$$

for some $g, h \in \mathcal{N}$, $g \hat{+} h \in \mathcal{N} \cup \{\hat{0}\}$. From this, a contradiction to π being a minimal element of the cone $\mathcal{D}_0(\mathcal{G}, b)$ is reached, in much the same way as in the proof of Theorem 4.

REFERENCES

[1] J. ARAOZ, *Polyhedral neopolarities,* Ph.D. thesis, Dept. of Computer Sciences and Applied Analysis, Univ. of Waterloo, Waterloo, Ontario, December 1973.

[2] C. A. BURDET AND E. L. JOHNSON, *A subadditive approach to solve linear integer programs,* Ann. Discrete Math., 1 (1977), pp. 117–143.

[3] R. E. GOMORY AND E. L. JOHNSON, *Some continuous functions related to corner polyhedra,* Math. Programming, 3 (1972), pp. 23–85.

[4] R. G. JEROSLOW, *The principles of cutting plane theory,* Carnegie Mellon University, Pittsburgh, PA, 1974.

[5] E. L. JOHNSON, *Support functions, blocking pairs, and anti-blocking pairs,* Math. Programming Stud., 8 (1978), pp. 167–196.

CHAPTER IX

Subadditive Characterizations

1. The blocker $\mathcal{B}(\mathcal{G}, b)$. In this chapter, we only consider master problems. First, vertices of the blocker $\mathcal{B}(\mathcal{G}, b)$ will be characterized. We need only be concerned with valid inequalities since $\mathcal{B}(\mathcal{G}, b)$ has no lineality and has extreme rays given by unit vectors. The nonnegative valid inequalities which are facets of $\mathcal{P}(\mathcal{G}, b)$ are precisely the vertices of $\mathcal{B}(\mathcal{G}, b)$. We characterize those vertices.

THEOREM 1. *The vertices of $\mathcal{B}(\mathcal{G}, b)$ are the vertices of $\overline{\mathcal{F}}_+(\mathcal{G}, b)$, the set of nonnegative subadditive valid inequalities, which are minimal.*

Before proving the theorem, let us characterize the minimal vertices of $\overline{\mathcal{F}}_+(\mathcal{G}, b)$. We claim they are the vertices of

(1) $\quad \pi(g) \geq 0, \quad g \in \mathcal{N},$

(2) $\quad \pi(g \hat{+} h) \leq \pi(g) + \pi(h), \quad g, h, g + h \in \mathcal{N},$

(3) $\quad 1 \leq \pi(b)$

which satisfy:

(4) $\quad$ for each $g \in \mathcal{N}$, either $g = b$ and $\pi(g) = 1$ or there exists $h \in \mathcal{N}$ such that $g \hat{+} h = b$ and $\pi(b) = \pi(g) + \pi(h)$.

THEOREM 2. *The minimal nonnegative valid inequalities are the ones for which (4) holds.*

Proof. If π is a nonnegative valid inequality for which (4) holds, then π is minimal because $\pi(g)$ cannot be made smaller without violating $-\pi \cdot t \geq 1$ for t given by $t(g) = t(h) = 1$ and the other t's equal to zero.

Conversely, suppose π is a minimal nonnegative valid inequality and (4) is violated for some $g \in \mathcal{N}$. Then, we first show that (4) is violated for some $g \in \mathcal{N}$ having $\pi(g) > 0$. If $\pi(g) = 0$ and (4) is violated for g, then for some $\delta > 0$,

$$\pi(h) \geq 1 + \delta \quad \text{for all } h \in \mathcal{N} \text{ such that } b = g \hat{+} h.$$

Clearly, $\pi(h) > 0$ for any such h, so if (4) is violated then we have our $\pi(h) > 0$ case. Otherwise, for any such h, there is some $\hat{h} \in \mathcal{N}$ such that

$$\pi(h) + \pi(\hat{h}) = 1.$$

By $\pi(h) \geq 1 + \delta$,

$$1 = \pi(h) + \pi(\hat{h}) \geq 1 + \delta + \pi(\hat{h}), \quad \text{or}$$

$$\pi(\hat{h}) \leq -\delta < 0,$$

contradicting $\pi(\hat{h}) \geq 0$.

To return to the main line of proof, let (4) be violated for some h with $\pi(h) > 0$. Then, for some $\delta > 0$,

$$\pi(h) + \pi(\hat{h}) \geq 1 + \delta, \quad \text{all } \hat{h} \text{ such that } h \stackrel{+}{=} \hat{h} = b.$$

Define $\rho(g)$, $g \in \mathcal{N}$, by

$$\rho(g) = \begin{cases} \dfrac{1}{1+\delta}\pi(h), & g = h, \\ \pi(g), & \text{otherwise.} \end{cases}$$

By $\pi(h) > 0$ and $\delta > 0$, $0 < \rho(h)$. A contradiction will be reached if ρ is valid. For any solution t,

$$\sum_{g \in \mathcal{N}} \rho(g)t(g) = \sum_{\substack{g \in \mathcal{N} \\ g \neq h}} \pi(g)t(g) + \frac{1}{1+\delta}\pi(h)t(h).$$

If $t(h) \geq (1+\delta)/\pi(h)$, then clearly $\rho \cdot t \geq 1$. If $t(h) = 0$, then $\rho \cdot t = \pi \cdot t \geq 1$. Otherwise, $1 \leq t(h) < (1+\delta)/\pi(h)$. Then

$$\sum_{g \in \mathcal{N}} \rho(g)t(g) = \sum_{\substack{g \in \mathcal{N} \\ g \neq h}} \pi(g)t(g) + \pi(h)t(h) - \frac{\delta}{1+\delta}\pi(h)t(h)$$

$$= \left(\sum_{\substack{g \in \mathcal{N} \\ g \neq h}} \pi(g)t(g) + \pi(h)(t(h) - 1) \right) + \pi(h) - \frac{\delta}{1+\delta}\pi(h)t(h)$$

$$\geq \pi\left(\sum_{\substack{g \in \mathcal{N} \\ g \neq h}} gt(g) \stackrel{+}{=} h(t(h) - 1) \right) + \pi(h) - \delta,$$

by subadditivity of minimal valid inequalities and use of $t(h) < (1+\delta)/\pi(h)$. Let

$$\hat{h} = \sum_{\substack{g \in \mathcal{N} \\ g \neq h}} gt(g) \stackrel{+}{=} h(t(h) - 1).$$

Then, $\hat{h} \stackrel{+}{=} h = \sum gt(g) = b$, so

$$\pi(h) + \pi(\hat{h}) \geq 1 + \delta,$$

and $\rho \cdot t \geq 1$, which completes the proof of Theorem 2.

Proof of Theorem 1. We have already shown (Theorem 4 of Chapter VIII) that minimal valid inequalities are subadditive. It is also clear that the vertices of $\mathcal{B}(\mathcal{G}, b)$ are minimal (see, e.g., Fulkerson [1] or Gomory and Johnson [2, Thm. 1.1]). Hence, every vertex of $\mathcal{B}(\mathcal{G}, b)$ is in $\overline{\mathcal{S}}_+(\mathcal{G}, b)$ and so must be a vertex of $\overline{\mathcal{S}}_+(\mathcal{G}, b)$, since $\overline{\mathcal{S}}_+(\mathcal{G}, b) \subseteq \mathcal{B}(\mathcal{G}, b)$. Every vertex of $\mathcal{B}(\mathcal{G}, b)$ is minimal. It only remains to show that every vertex π of $\overline{\mathcal{S}}_+(\mathcal{G}, b)$ which is minimal is also a vertex of $\mathcal{B}(\mathcal{G}, b)$. If it were not, then π would be a midpoint of two different nonnegative valid inequalities. If these two valid inequalities were minimal, then they would be subadditive, contradicting the original assumption that π is a vertex of $\overline{\mathcal{S}}_+(\mathcal{G}, b)$. Thus, the lemma below completes the proof.

LEMMA 3. *If π is a minimal valid inequality and $\pi = \frac{1}{2}\rho^1 + \frac{1}{2}\rho^2$ for $\rho^1 \neq \rho^2$ and ρ^1, ρ^2 valid inequalities, then ρ^1 and ρ^2 must also be minimal.*

Proof. Suppose not. Then, say, ρ^1, is not minimal so there is a $\rho^3 < \rho^1$ where ρ^3 is also a valid inequality. But

$$\pi > \tfrac{1}{2}\rho^3 + \tfrac{1}{2}\rho^2,$$

contradicting π being minimal.

2. Valid inequalities. The polyhedron $\mathcal{Q}(\mathcal{G}, b)$ is defined by

(5) $$\sum_{g \in \mathcal{N}} \pi(g) t(g) \geq 1,$$

(6) $$\sum_{g \in \mathcal{N}} \rho(g) t(g) \geq 0,$$

(7) $$\sum_{g \in \mathcal{N}} \sigma(g) t(g) \geq -1,$$

where (5) is taken over the vertices π of $\mathcal{D}_+(\mathcal{G}, b)$, (6) is taken over the extreme rays ρ of $\mathcal{D}_0(\mathcal{G}, b)$, and (7) is taken over the vertices σ of $\mathcal{D}_-(\mathcal{G}, b)$.

THEOREM 4. *The valid equations are equal to the lineality of $\mathcal{S}(\mathcal{G})$. The extreme rays of $\mathcal{D}_0(\mathcal{G}, b)$ are a subset of the unit vectors together with the extreme rays of $\mathcal{S}_0(\mathcal{G}, b)$ which are minimal. The vertices of $\mathcal{D}_+(\mathcal{G}, b)$ ($\mathcal{D}_-(\mathcal{G}, b)$) are the vertices of $\mathcal{S}_+(\mathcal{G}, b)$ ($\mathcal{S}_-(\mathcal{G}, b)$) which are minimal.*

Proof. The part on the lineality has already been shown (Theorems 1 and 5 of Chapter VIII). We also know by Theorem 2 of Chapter VIII that the vertices of

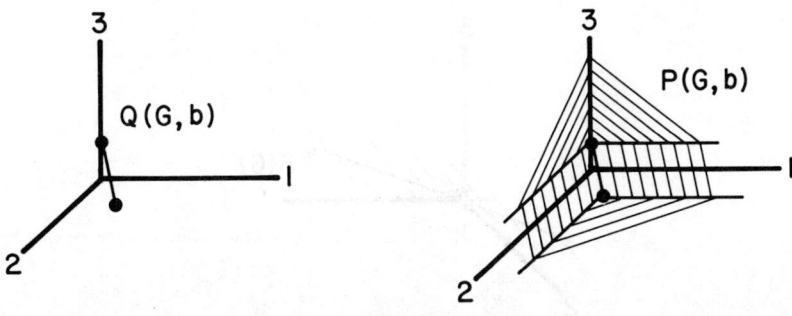

FIG. 1

$\mathscr{S}_-(\mathscr{G}, b)$ $(\mathscr{S}_-(\mathscr{G}, b))$ are in $\mathscr{D}_+(\mathscr{G}, b)$ $(\mathscr{S}_-(\mathscr{G}, b))$. The part of the theorem for $\mathscr{D}_0(\mathscr{G}, b)$ and $\mathscr{S}_0(\mathscr{G}, b)$ is Theorem 6 of Chapter VIII.

The line of proof for $\mathscr{D}_+(\mathscr{G}, b)$ and $\mathscr{D}_-(\mathscr{G}, b)$ is similar to that for $\mathscr{D}_0(\mathscr{G}, b)$ and for $\mathscr{B}(\mathscr{G}, b)$ and is essentially the same as in [2]. Minimal elements of $\mathscr{D}_+(\mathscr{G}, b)$ $(\mathscr{D}_-(\mathscr{G}, b))$ are in $\mathscr{S}_+(\mathscr{G}, b)$ $(\mathscr{S}_-(\mathscr{G}, b))$. Extreme points of $\mathscr{D}_+(\mathscr{G}, b)$ $(\mathscr{D}_-(\mathscr{G}, b))$ are minimal and, hence, in $\mathscr{S}_+(\mathscr{G}, b)$ $(\mathscr{S}_-(\mathscr{G}, b))$. Conversely, any extreme point in $\mathscr{S}_+(\mathscr{G}, b)$ $(\mathscr{S}_-(\mathscr{G}, b))$ which is also minimal must also be extreme in $\mathscr{D}_+(\mathscr{G}, b)$ $(\mathscr{D}_-(\mathscr{G}, b))$ because otherwise it would be the midpoint of two other valid inequalities which must be minimal (Lemma 3) and, hence, subadditive (Theorem 4 of Chapter VIII), contradicting extremality in $\mathscr{S}_+(\mathscr{G}, b)$ $(\mathscr{S}_-(\mathscr{G}, b))$.

Example 1. Consider the semigroup $\mathscr{G}$ having addition table

	0	1	2	3	∞
0	0	1	2	3	∞
1	1	∞	3	∞	∞
2	2	3	∞	∞	∞
3	3	∞	∞	∞	∞
∞	∞	∞	∞	∞	∞

Let $b = 3$. Then, the solutions are $(1, 1, 0)$ and $(0, 0, 1)$. The polyhedra $\mathscr{Q}(\mathscr{G}, b)$ and $\mathscr{P}(\mathscr{G}, b)$ are shown in Fig. 1. The cone $\mathscr{S}(\mathscr{G})$ is just the half space $\pi_3 \leq \pi_1 + \pi_2$, as shown in Fig. 2. Intersecting it with $\pi_3 \geq 0$ gives the cone: $\pi_3 \geq 0$, $\pi_1 + \pi_2 \geq 0$, i.e., $\mathscr{S}_0(\mathscr{G}, b)$. The polyhedron $\mathscr{S}_+(\mathscr{G}, b)$ has lineality the same as $\mathscr{S}_0(\mathscr{G}, b)$, i.e., the line $\pi_2 = -\pi_1$, $\pi_3 = 0$. We can generate $\mathscr{S}_+(\mathscr{G}, b)$ from any point on the line $\pi_3 = 1$, $\pi_1 + \pi_2 = 1$, e.g., $(1, 0, 1)$, and $\mathscr{S}_0(\mathscr{G}, b)$.

Figure 3 shows $\mathscr{D}_0(\mathscr{G}, b)$ and $\mathscr{D}_+(\mathscr{G}, b)$. We get $\mathscr{B}(\mathscr{G}, b)$ from the intersection of $\mathscr{D}_+(\mathscr{G}, b)$ with $\mathscr{R}_+^3$. The system of inequalities for $\mathscr{P}(\mathscr{G}, b)$ is

$$t_1 + t_3 \geq 1,$$
$$t_2 + t_3 \geq 1,$$
$$t_1, t_2, t_3 \geq 0$$

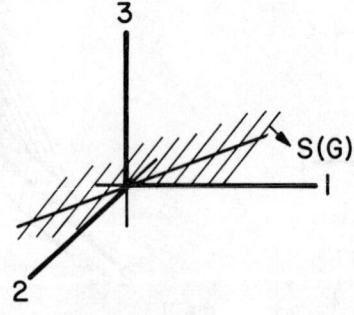

Fig. 2

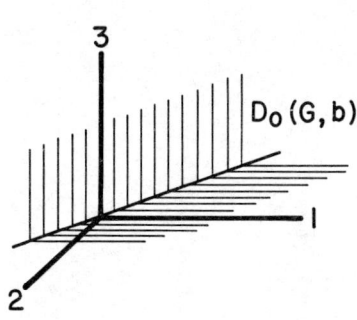 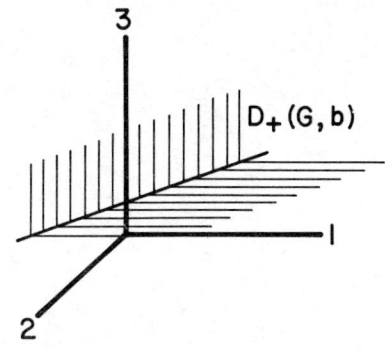

FIG. 3

and the system of inequalities for $\mathcal{B}(\mathcal{G}, b)$ is

$$\pi_3 \geqq 1,$$
$$\pi_1 + \pi_2 \geqq 1,$$
$$\pi_1, \pi_2, \pi_3 \geqq 0.$$

For $\mathcal{Q}(\mathcal{G}, b)$ there are two equations

$$t_1 - t_2 = 0,$$
$$t_1 + t_3 = 1.$$

These constitute a basis for the lineality of $\mathcal{S}(\mathcal{G})$. They turn out to be sufficient for defining $\mathcal{Q}(\mathcal{G}, b)$.

REFERENCES

[1] D. R. FULKERSON, *Blocking polyhedra*, Graph Theory and its Applications, B. Harris, ed., Academic Press, New York, 1970, pp. 93–112.
[2] R. E. GOMORY AND E. L. JOHNSON, *Some continuous functions related to corner polyhedra*, Math. Programming, 3 (1972), pp. 23–85.

CHAPTER X

Duality

1. A dual problem for master problems. The master semigroup problem is equivalent to

(1) $\quad t(g) \geq 0, \quad g \in \mathcal{N},$

(2) $\quad \sum_{g \in \mathcal{N}} \pi^k(g)t(g) \geq \pi^k(b), \quad \pi^k$ an extreme ray of $\mathcal{S}(\mathcal{G}),$

(3) $\quad \text{minimize } c(g)t(g),$

in the sense that there is an optimum solution to the linear program (1), (2), (3) which is a solution to the master semigroup problem.

Ordinary linear programming gives us a dual problem:

(4) $\quad \lambda_k \geq 0;$

(5) $\quad \sum_k \lambda_k \pi^k(g) \leq c(g);$

(6) $\quad \text{maximize } \sum_k \lambda_k \pi^k(b).$

For λ satisfying (4) and (5), $\sum \lambda_k \pi^k / \sum \lambda_k$ is also in $\mathcal{S}(\mathcal{G})$, and so we can state a weaker dual problem:

(7) $\quad \pi(g) \leq c(g), \quad g \in \mathcal{N};$

(8) $\quad \text{maximize } \pi(b)$

over all π satisfying

(9) $\quad \pi(g \hat{+} h) \leq \pi(g) + \pi(h), \quad g, h, g \hat{+} h \in \mathcal{N},$

(10) $\quad 0 \leq \pi(g) + \pi(g) \quad \text{if } g, h \in \mathcal{N} \text{ and } g \hat{+} h = \hat{0}.$

If we had a linear description of minimal inequalities, that, too, could be imposed on π.

In this form, weak duality is simple to prove: if t satisfies (1) and (2) and if π satisfies (7), (9), and (10), then

$$\pi(b) = \pi\left(\sum_{g \in \mathcal{N}} gt(g)\right)$$
$$\leq \sum_{g \in \mathcal{N}} \pi(g)t(g)$$
$$\leq \sum_{g \in \mathcal{N}} c(g)t(g).$$

Strong duality is shown from linear programming duality by the fact that each inequality (2) is valid.

In order for $\pi(b) = ct$ to hold, it is necessary that

(11) $\quad\quad\quad\quad t(g) \geq 1 \quad \text{implies} \quad \pi(g) = c(g), \quad \text{and}$

(12) $\quad\quad\quad\quad$ if $h \in \mathcal{N}$, $t(h) \geq 1$, and $h' = \sum gt'(g)$ for

$\quad\quad\quad\quad 0 \leq t'(g) \leq t(g)$, $t'(g)$ integer, and $t'(h) < t(h)$,

$\quad\quad\quad\quad$ then $\pi(h \,\hat{+}\, h') = \pi(h) + \pi(h')$.

Condition (11) comes from $\sum \pi(g)t(g) = \sum c(g)t(g)$ and is the *complimentary slackness condition* of linear programming. Condition (12) comes from

$$\pi\left(\sum gt(g)\right) = \sum \pi(g)t(g),$$

along with (9), and we call (12) the *complimentary linearity condition*. In fact, these two conditions are not only necessary, but sufficient to assure equality of the objectives.

In stating a dual problem, a stronger version is one which restricts the dual function π as much as possible. The hard part of the duality theorem is to show that equality of the objectives can be achieved. The more restricted π is the harder it is to show this equality and the stronger is the result. When we have a linear characterization of the minimal valid inequalities:

(13) $\quad\quad\quad\quad\quad\quad \pi(b) = \pi(h) + \pi(\hat{h})$

for pairs $h, \hat{h}$ satisfying $b = h + \hat{h}$, then imposing (13) along with (9) and (10) is as restricted as we can make the dual because π is then restricted to lie in the cone generated by the vertices of $\mathcal{B}(\mathcal{G}, b)$.

This duality was explicitly stated and used for the cyclic group problem in [5 Thm. 2]. For other references, see Burdet and Johnson [3] and Jeroslow [4].

2. Semigroup problems for general $\mathcal{N}$. For this section, let $\mathcal{N} \subseteq \mathcal{G}$ be arbitrary, except, as always, $\hat{0} \notin \mathcal{N}$, $\infty \notin \mathcal{N}$, and $\mathcal{G}$ is the semigroup generated by $\mathcal{N}$. The semigroup problem is still equivalent to (1), (2), and (3) except that now some of the inequalities (2) may be redundant in the sense that they are implied by the others. For arbitrary problems, we make no effort to determine which are needed.

The same dual formulation holds here except that π is defined on all of $\mathcal{G} - \{\infty\}$, with $\pi(\hat{0}) = 0$, but is constrained by (7) only for $g \in \mathcal{N}$. That is, (9) and (10) must hold for $\mathcal{G}$, but (7) is only for $\mathcal{N}$.

We cannot say when this dual problem is as strong as possible because we do not make any effort to know exactly the faces for arbitrary $\mathcal{N} \subseteq \mathcal{G}$. However, imposing minimality where possible is as strong as the dual can be made so that it will work for all $\mathcal{N} \subseteq \mathcal{G}$. In general, this approach will be taken.

Example 1 (Araoz [1]). Let us take the set covering example. Let $\mathcal{N}$ be a family of subsets $S \subseteq X$. Then, $\mathcal{G}$ is the closure under union of $\mathcal{N}$ with $\hat{+}$ being union. The zero of $\mathcal{G}$ is the empty set. In this case, the integer variables will take on 0 1 values because $\mathscr{G}$ is idempotent: $S \cup S = S$. The problem, with $b = X$, is

$$\bigcup_{S \in \mathscr{N}} St(s) = X,$$

and minimize $\sum_{S \in \mathscr{N}} c(S) t(S)$,

where $c(S) \geq 0$ is the objective coefficient for S. Here, $t(s) = 0$ or 1 depending on whether S is used to cover X.

Subadditivity means

$$\pi(S \cup T) \leq \pi(S) + \pi(T), \qquad S, T \in \mathscr{G}.$$

In this case, $\mathscr{G}$ has no ∞ element.

Minimality for π is equivalent to

$$\pi(X) = \pi(S) + \pi(X - S), \qquad S \in \mathscr{G},$$

provided all $X - S \in \mathscr{G}$ when $S \in \mathscr{G}$. To assure this condition, we could take $\mathscr{G}$ to be the closure of $\mathscr{N}$ under union and complementation. The dual problem is to find a subadditive, minimal π such that

$$\pi(S) \leq c(S), \qquad S \in \mathscr{N},$$

so as to maximize $\pi(X)$. Araoz [1] shows that the π here can also be assumed to be monotone increasing.

Example 2. Let us now take set partitioning as the problem. As in Example 1, let $\mathscr{N}$ be a family of subsets of X. Define $\mathscr{G}$ to be the closure under disjoint union of $\mathscr{N}$ with $\hat{+}$ taken as disjoint union:

$$S \hat{+} T = \begin{cases} S \cup T & \text{if } S \cap T = \emptyset, \\ \infty & \text{otherwise.} \end{cases}$$

As before $\hat{0} = \emptyset$ is the empty set.

Subadditivity is now only required for disjoint S and T:

$$\pi(S \cup T) \leq \pi(S) + \pi(T) \quad \text{if } S \cap T = \emptyset.$$

Minimality is equivalent to

$$\pi(X) = \pi(S) + \pi(X - S), \qquad S \in \mathscr{G}$$

provided all complements $X - S$ belong to $\mathscr{G}$.

The dual problem is to maximize $\pi(X)$ subject to subadditivity and

$$\pi(S) \leq c(S), \qquad S \in \mathscr{N}.$$

3. Lifting procedures. This type of procedure was introduced in [5] and is used by Burdet and Johnson [2], [3]. The particular algorithm given here is that found in [6]. The problem considered is

$$t(j) \geq 0, \quad j \in \mathcal{N},$$

$$\sum_{j \in \mathcal{N}} jt(j) \geq b,$$

$$\text{minimize} \sum c(j)t(j) \quad \text{where } c(j) \geq 0.$$

The lifting procedure has an initializing step and two main iterative steps.

INITIAL: Let $H = \{0\}$ be the set of *hit points*, $L = \{b\}$ be the set of *lifting points*, and $C = \{(j, c(j)) \mid j \in \mathcal{N}\}$ be the *constraining set*. Let $\pi(0) = 0$ and $\pi(b) = 0$.

LIFT: Form the piecewise linear function π going through the points $(h, \pi(h))$, $h \in H$, and $(l, \pi(l))$, $l \in L$, and interpolating linearly in between. Now, increase each π_l, $l \in L$, by the same amount $\varepsilon \geq 0$ until some $(j, c(j)) \in C$ lies on the function, i.e., $\pi(j) = c(j)$. If $j \in L$, terminate. Otherwise, go to HIT.

HIT: Put j in H and fix $\pi(j) = c(j)$. Remove $(j, c(j))$ from C. Put $b - j$ in L with $\pi(b - j) = \pi(b) - \pi(j)$. Put in C every $(j \hat{+} h, \pi(j) + \pi(h))$ for $h \in H$, where $j \hat{+} h = b$ if $j + h > b$. Return to LIFT.

Some refinements in [3] could be used here, but do not change the overall scheme.

The function π generated will be complementary, $\pi(b) = \pi(j) + \pi(b - j)$, hence, minimal.

Example 3. Consider the problem of minimizing z subject to

$$2t(2) + 5t(5) + 6t(6) \geq 8,$$

$$t(j) \geq 0 \text{ and integer}, \quad j = 2, 5, 6,$$

$$2t(2) + 3t(5) + 4t(6) = z.$$

The first stage produces the function shown in Fig. 1. Hitting $(5, 3)$ generates the new point $(8, 6)$ in C. Now, $H = \{0, 5\}$, $L = \{8, 3\}$, and $C = \{(2, 2), (6, 4), (8, 6)\}$.

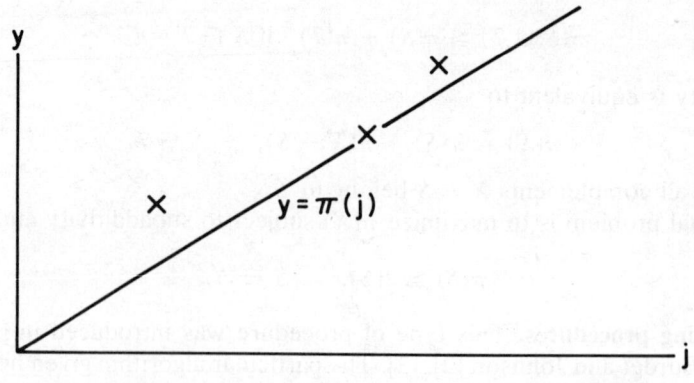

Fig. 1

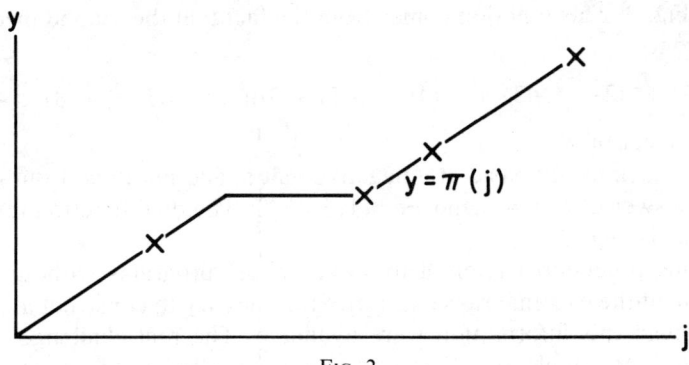

Fig. 2

The next lifting process results in the function shown in Fig. 2. The points (2, 2), (6, 4), and (8, 6) in C are all hit. Since one of them is in L, namely 8, the problem is solved. The optimal primal is to use $t(5)$ twice. The optimal dual is the function shown in Fig. 2.

Let us now compare this solution with solving the linear programming problem over the facets of the problem. By direct inspection, the linear program needed is

$$t(2), t(5), t(6) \geq 0,$$

$$t(2) + t(5) + t(6) \geq 2,$$

$$t(2) + 2t(5) + 3t(6) \geq 4,$$

$$2t(2) + 3t(5) + 4t(6) = z.$$

Using the dual simplex, we begin with both slacks basic, at -2 and -4, and a dual solution of $\pi_1 = 0$, $\pi_2 = 0$. Choosing the second slack to eliminate gives primal and dual solutions of $t(6) = 1\frac{1}{3}$, $\pi_2 = 1\frac{1}{3}$. The dual function generated is

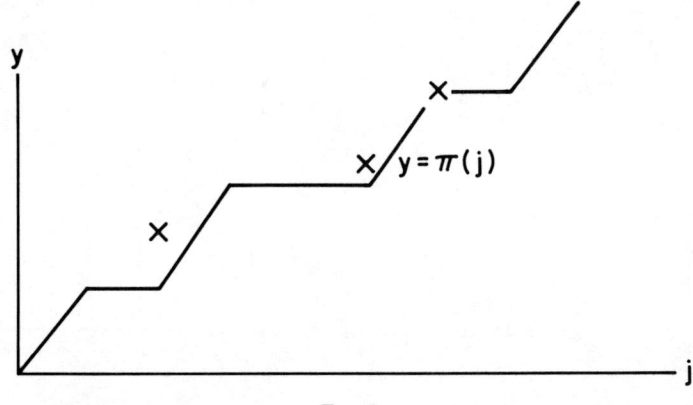

Fig. 3

shown in Fig. 3. This function comes from the fact that the second inequality is derived from

$$t(1) + t(2) + 2t(3) + 2t(4) + 2t(5) + 3t(6) + 3t(7) + 4t(8) \geqq 4,$$

which is a facet of $\mathcal{K}^{\geqq}(8)$.

The first slack is still basic at a negative value, and iterating it out gives the optimum answer of $t(5) = 2$ and $\pi_1 = 1$, $\pi_2 = 1$. The dual function resulting is the same as in Fig. 2.

The lifting procedures attempt to solve integer programs without knowing facets or all of the extreme rays of $\mathcal{S}(\mathcal{G})$. Still, they try to construct a dual function as though this information were available. The real challenge is to find appropriate function classes in order to carry out efficient lifting procedures.

REFERENCES

[1] J. ARAOZ, *Polyhedral neopolarities*, Ph.D. thesis, Dept. of Computer Sciences and Applied Analysis, Univ. of Waterloo, Waterloo, Ontario, December 1973.
[2] C. A. BURDET AND E. L. JOHNSON, *A subadditive approach to the group problem of integer programming*, Math. Programming Stud., 2 (1974), pp. 51–71.
[3] ———, *A subadditive approach to solve integer programs*, Ann. Discrete Math., (1977), pp. 117–143.
[4] R. JEROSLOW, *The principles of cutting plane theory, Part II: Algebraic methods, disjunctive methods*, Management Sci. Rep. 370 (revised), Carnegie–Mellon University, Pittsburgh, PA, September 1975.
[5] E. L. JOHNSON, *Cyclic groups, cutting planes, and shortest paths*, Mathematical Programming, T. C. Hu and S. Robinson, eds. Academic Press, New York, 1973.
[6] ———, *On the group problem and a subadditive approach to integer programming*, Annals of Discrete Mathematics 5: Discrete Optimization II, P. L. Hammer et al., eds, North-Holland, Amsterdam, 1979, pp. 97–112.